Advances in Biogas Desulfurization

Advances in Biogas Desulfurization

Special Issue Editor

Martín Ramírez

MDPI • Basel • Beijing • Wuhan • Barcelona • Belgrade • Manchester • Tokyo • Cluj • Tianjin

Special Issue Editor
Martín Ramírez
University of Cádiz
Spain

Editorial Office
MDPI
St. Alban-Anlage 66
4052 Basel, Switzerland

This is a reprint of articles from the Special Issue published online in the open access journal *ChemEngineering* (ISSN 2305-7084) (available at: https://www.mdpi.com/journal/ChemEngineering/special_issues/Biogas_Desulfurization).

For citation purposes, cite each article independently as indicated on the article page online and as indicated below:

LastName, A.A.; LastName, B.B.; LastName, C.C. Article Title. *Journal Name* **Year**, *Article Number*, Page Range.

ISBN 978-3-03928-660-7 (Pbk)
ISBN 978-3-03928-661-4 (PDF)

Contents

About the Special Issue Editor

Martín Ramírez (Ph.D.) has been an Associate Professor in the Faculty of Science of the University of Cadiz (UCA, Spain) since 2017. He obtained a B.S. in Chemical Engineering in 2002 and M.S. and Ph.D. degrees in the Chemical Science and Technology program at the same institution in 2007. He was the head of the chemical engineering section in a technology driven spin-off for 3 years (2008–2011). His research activities have mainly focused on effluent gases biofiltration, such as air (odor removal), biogas (desulfurization and upgrading), and bioremediation of high value metals. Currently, he is a partner team leader of a LIFE EU project (Biogasnet) and project leader of two projects related to the bioremediation of platinum group metals from three-way catalysts. He is co-author of 28 scientific contributions to ISI journals (H-index of 14, Scopus databases), 2 patents, 7 books chapters, more than 40 conference proceedings in international congresses; he has supervised five M.S. students, four Ph.D. theses, and four Ph.D. theses are in progress. M.R. belongs to the Editorial Board of *ChemEngineering* and serves as Guest Editor for *ChemEnginering* in this Special Issue "Advances in Biogas Desulfurization". He is regularly involved in the peer-review in ISI journals (more than 100 since 2009) and he has participated in the peer-reviewing processes of technical and scientific projects (NCSTE of Kazakahstan and UEFISCDI of Romania).

Preface to "Advances in Biogas Desulfurization"

The environmental impacts of non-renewable energies and crude oil depletion have increased the use of biogas. Biogas is a renewable energy source produced under anaerobic conditions by the degradation of organic matter. However, before its valorization, it needs to be desulfurized (H_2S removal) and/or upgraded (CO_2 removal). The main biogas uses are heat and power production, injection into the natural gas grid, fuel for solid oxide fuel cells, biogas reforming, and vehicle fuel. In all these applications, biogas needs to be desulfurized because H_2S causes corrosion and sulfur oxides emissions during biogas combustion.

In the last 15 years, the number of desulfurization technologies have increased and their performance has improved. However, shortcomings remain, such as elemental sulfur accumulation avoidance (packed bed bioreactors) and its separation (suspension biomass bioreactors). In addition, these technologies need to be extended to demonstration and industrial scales. This Special Issue shows some advances in biogas desulfurization.

Velasco et al. (Desulfurization of Biogas from a Closed Landfill under Acidic Conditions Deploying an Iron-Redox Biological Process) removed H_2S using a system composed of an absorption bubble column and a biotrickling filter.

Almenglo et al. (Application of Response Surface Methodology for H_2S Removal from Biogas by a Pilot Anoxic Biotrickling Filter) studied and modeled H_2S removal using an anoxic biotrickling filter.

Tayar et al. (Evaluation of Biogas Biodesulfurization Using Different Packing Materials) researched the effect of four packing materials for an anoxic biotrickling filter.

Okoro and Sun (A Systematic Qualitative and Economic-Based Quantitative Review of Alternative Strategies) reviewed the current state of biogas desulfurization technologies and the annual operation and annualized capital cost per unit of volume.

Le Borgne and Baquerizo (Microbial Ecology of Biofiltration Units Used for the Desulfurization of Biogas) reviewed the microbial ecology of biofiltration units. The main characteristics of sulfur-oxidizing chemotrophic bacteria are presented.

As the Guest Editor, I hope that the readership will find the information published in this Special Issue about biogas desulfurization useful. Finally, I would like to thank the effort contributed by all the authors, the publisher and the reviewers that allowed this book to be published.

Martín Ramírez
Special Issue Editor

Editorial

Special Issue "Advances in Biogas Desulfurization"

Martín Ramírez

Department of Chemical Engineering and Food Technologies, Wine and Agrifood Research Institute (IVAGRO), Faculty of Sciences, University of Cadiz, Puerto Real, 11510 Cádiz, Spain; martin.ramirez@uca.es; Tel.: +34-956-01-6286

Received: 5 March 2020; Accepted: 5 March 2020; Published: 9 March 2020

Abstract: This Special Issue contains three articles and two reviews. The biological reactors used in the studies were fed with real biogas from Landfill or STPs. One research article concerns the use of a pilot scale plant with a combined process with a chemical and biological system. The other two studies concern anoxic biotrickling filters, with one study focused on the study of variable operation and its optimization through the response surface methodology, and the other focused on the selection of packing material. The reviews concern the current state of biogas desulfurization technologies, including an economic analysis, and the microbial ecology in biofiltration units. This Issue highlights some of the most relevant aspects about biogas desulfurization.

Keywords: hydrogen sulfide; biogas; desulfurization; biotrickling filter; anoxic; response surface methodology; microbial ecology; sulfur-oxidizing bacteria; packing material; anaerobic digestion

---

1. Introduction

This Special Issue contains the invited submissions to a Special Issue of *ChemEngineering* on the topic "Advanced Biogas Desulfurization" [1–5]. Three research articles [1–3] and two reviews [4,5] have been published. Biogas is a renewable energy source produced by the biodegradation of organic matter under anaerobic conditions. The use of renewable energies is increasing due to global warming and the increasing price of fossil fuels. However, biogas needs to be desulfurized prior to use. The biogas composition mainly depends on the feedstock (sludge from sewage treatment plants (STPs), waste from the agri-food industry, the organic fraction of municipal solid waste, livestock manure, etc.), with the main components being methane (45%–75%) and carbon dioxide (20%–50%). However, hydrogen sulfide (H_2S) leads to corrosion and the combustion of non-desulfurized biogas produces the emission of SOx in flue gases. The H_2S concentration can range from a few ppmv (0.5–700 ppmv in landfill gas) up to more than 30,000 ppmv in the plant's pulp-making process; and biogas flow rates can be in the range from several hundred cubic meters per hour (usually in STPs) to several thousand cubic meters per hour (usually in landfills). The applications of biogas are also wide-ranging, with the most common being burning in motors to produce electricity or electricity and heat. However, biogas can also be used as a fuel for solid oxide fuel cells or for hydrogen production by biogas reforming. Moreover, in the case of upgrading (CO_2 removal) biogas can be injected into the natural gas grid or used as fuel for vehicles. This wide range of biogas flow rates, H_2S concentrations, types of applications and purification requirements, as well as biogas sources, results in a wide variety of technologies, which can be subdivided into those that involve physicochemical phenomena and those that involve biological processes.

This Special Issue aims to bring together the scientific/technical advances on physicochemical and/or biological processes for biogas desulfurization. Biogas desulfurization is considered to be essential by many stakeholders (biogas producers, suppliers of biogas upgrading devices, gas traders, researchers, etc.) around the world, as the importance of biogas desulfurization to allow its valorization is well understood.

2. Brief Overview of the Contributions to This Special Issue

Velasco et al. [1] carried out the desulfurization of landfill biogas ('Prados de la Montaña', Mexico) at the pilot scale by a combined process involving chemical and biological treatments. In this study, an Absorption Bubble Column (ABC) was used in which H_2S was absorbed and oxidized to elemental sulfur by ferric sulfate. A biotrickling filter (BTF) was employed for ferric sulfate regeneration (oxidation of Fe^{2+} produced in the ABC to Fe^{3+}) by an enriched acidophilic mineral-oxidizing bacteria consortium (AMOB). The first reported application of an iron-based process and biological regeneration was reported in 1984 by Barium Chemical Ltd. and this approach is known as the Bio-SR process using *Acidithiobacillus ferrooxidans* and a jet-scrubber for H_2S oxidation. Since then, numerous configurations have been published, but the main drawbacks are related to the elemental sulfur separation and jarosite formation. Jarosite helps to develop the biofilm growth but it reduces the amount of Fe^{3+} and, therefore, its formation must be controlled, usually by controlling the pH. The effect of no pH control and no forced convection of air on the iron oxidation rates and H_2S removal were studied. In this study, removal efficiencies (REs) higher than 99.5% were achieved for H_2S concentrations in the range 120–250 ppmv (Empty Bed Residence Time (EBRT) of 4.5 min in the ABC). The system was successfully operated for around seven months with minimal energy input and a metastable operation the zone between $2FeOH^{2+}$ and Fe^{2+}. However, elemental sulfur was accumulated on the packed bed of the BTF and this was not recovered in the settler due to its colloidal nature.

Almenglo et al. [2] studied the H_2S removal from biogas produced in an STP ('Bahía Gaditana', Cádiz, Spain) by an anoxic BTF at the pilot scale (packed bed volume of 0.167 m^3). The effect of the biogas flow rate, trickling liquid velocity (TLV) and nitrate concentration on the H_2S RE and elimination capacity (EC) were studied using a full factorial design (3^3). Anoxic biofiltration is a promising technology for biogas desulfurization because it avoids biogas dilution and reduces the risk of explosion when compared to aerobic BTFs. In fact, in the past seven years, there has been a significant increase in the number of published studies in this area. In this study, the highest H_2S RE values were obtained at a TLV of 15.27 m h^{-1}, with RE values of 99.53, 97.65 and 92.13% for EBRTs of 600, 200 and 120 s, respectively. Therefore, the maximum and critical ECs were 158.83 gS m^{-3} h^{-1} (RE 92.13%) and 34.93 gS m^{-3} h^{-1} (RE 99.53%), respectively. Higher values can be found in the literature for laboratory scale systems with a higher height:diameter ratio, but this pilot plant was one of the first anoxic BTFs to be installed in an STP, thus demonstrating the feasibility of this technology under real operating conditions (fluctuations in the biogas composition, weather, etc.). Moreover, experimental data were adjusted using Ottengraf's model, which allows the H_2S concentration along the bed to be predicted in a simple way. In contrast, dynamic models have been published previously by the same authors and this provided a better understanding of the process—although these models are more complex and are seldom used.

Tayar et al. [3] studied different packing materials for an anoxic BTF (packed bed volume of 3 L). One of the main drawbacks of BTFs, both aerobic and anoxic, is the clogging of the packed bed. Clogging can be caused by the accumulation of elemental sulfur and/or biomass growth. However, in most cases it is due to elemental sulfur accumulation, since the biofilm growth rate is low. It can be seen from the literature that elemental sulfur production can be controlled by increasing the electron acceptor feed (nitrate, nitrite or oxygen) but sulfur formation is unavoidable. In this respect, the selection of the packing material is critical as it will affect the amount of sulfur and biomass that can attach to the packing in the bed. In this study, four packing materials were tested: strips of polyvinyl chloride (PVC), polyethylene terephthalate (PET), polytetrafluoroethylene (Teflon®) and open-pore polyurethane foam (OPUF). PVC was chosen due to the high concentration of biomass, although it was lower than for OPUF, and its low cost. The BTF performance showed high H_2S removal (95.72%) and EC (98 gS m^{-3} h^{-1}), with values similar to those obtained in previous studies carried out with OPUF. Therefore, PVC could be a potential low-cost packing material for use in BTFs.

Okoro and Sun [4] submitted an interesting review about the current state of biogas desulfurization technologies: physicochemical, biological, in situ, and post-biogas desulfurization strategies. Moreover, a review of the annual operation and annualized capital cost per unit volume was carried out. To perform a cost analysis is a challenging undertaking and there are very few published studies. Biogas stakeholders could make decisions about the best technology based on these results. However, there are numerous factors that make the comparison complex: lack of data from companies, differences between studies (scale, source of biogas, etc.), location in different countries, supplies (prices of electricity, chemicals, etc.), etc. In this review, the authors carried out a thorough review; for instance, studies from member countries of the OECD and uncertainties about the 50%–150% variation in the cost were considered. The study shows that in situ chemical dosing is the cheapest biogas desulfurization technique, although limitations were identified in terms of the system control costs and environmental impact due to the continuous chemical supply. Moreover, the integration of several technologies could be of interest to reduce the weaknesses of each desulfurization strategy.

Le Borgne and Baquerizo [5] present a review on the microbial ecology of biofiltration units for biogas desulfurization. Moreover, a review of the biofiltration technologies is included: conventional biofilter, BTF and bioscrubbers. Biological H_2S oxidation can be carried out under aerobic or anoxic conditions. Therefore, the main chemotrophic Sulfur Oxidizing Bacteria (SOB) will depend on the final electron acceptor. The review shows the microbial ecology in aerobic and anoxic BTFs through molecular techniques such as fingerprint methods (PCR-DGGE, T-RFLP), fluorescence in situ hybridization (FISH), or next generation sequencing technologies (450-pyrosequencing or Illumina platforms). As one would expect, the environmental conditions had a direct impact on the diversity of bacterial communities and their structure and dynamics. However, not enough is currently known about the role of the main populations in these bioreactors.

3. Gaps in Biogas Desulfurization

In the past 15 years, there have been significant advances in the development of biological desulfurization processes. However, there are still shortcomings that require further investigation. To my mind, one of the main issues is the formation of elemental sulfur in bioreactors with packing material, such as BTFs, although there has been a significant advance in the increase of the oxygen mass transfer in aerobic bioreactors, such as aerobic BTFs. In the case of anoxic bioreactors, it is possible to feed the system with high nitrate or nitrite concentrations, although this entails a significant cost (environmental and economic) in terms of the use of chemical compounds. However, the feasibility of feeding nitrified effluent from ammonium-rich wastewater has been demonstrated, thus avoiding the above impact. In any case, elemental sulfur formation is unavoidable and further research is needed to prevent its accumulation. A possible solution would be the use of suspended biomass bioreactors, in which clogging would be avoided. In these bioreactors, a new issue would be the separation of elemental sulfur in an economic way.

Another important gap is to determine the role of the key populations to avoid operational outages in the bioreactors. Likewise, progress can be made in the control systems by improving the mathematical model using the latest advances in microsensors. A microsensor has been developed to measure pH and O_2 profiles in the biofilm and, in this respect, it would be interesting to develop new microsensors to measure sulfide, nitrate and nitrite profiles.

A high priority area is the scaling-up of desulfurization technologies to the demonstration or industrial scales. Many of these systems have only reached the pilot scale, so it is necessary to develop larger plants in order to obtain long-term operational data, to determine the operational limits and to evaluate economic and environmental impacts. In this regard, some studies have already been carried out, but they are very scarce, and greater effort is needed in this direction. In other cases, such as aerobic BTFs, all of this information is available and a better dissemination in companies is necessary in order to increase the number of industrial plants. All of this information will allow the evaluation of these technologies and enable the installation of new plants. It is also necessary to reduce the gas residence time in order to design smaller equipment, minimize energy consumption and integrate desulfurization systems in plants to avoid the consumption of chemical reagents. Finally, it would be interesting to look for new biological processes for biogas revalorization. For instance, other value-added products could be obtained from the methane and carbon dioxide present in the biogas that can be integrated into the biological desulfurization processes.

Acknowledgments: I would like to thank the publisher and the Editorial staff Team, especially Joqeen Meng, for inviting me to be Guest Editor of this Special Issue. Moreover, I would like to thank all of the contributors and reviewers for their efforts.

Conflicts of Interest: The author declares no conflict of interest

References

1. Velasco, A.; Franco-Morgado, M.; Revah, S.; Arellano-García, L.; Manzano-Zavala, M.; González-Sánchez, A. Desulfurization of Biogas from a Closed Landfill under Acidic Conditions Deploying an Iron-Redox Biological Process. *Chemengineering* **2019**, *3*, 71.

2. Almenglo, F.; Ramírez, M.; Cantero, D. Application of Response Surface Methodology for H_2S Removal from Biogas by a Pilot Anoxic Biotrickling Filter. *Chemengineering* **2019**, *3*, 66.

3. Tayar, S.; de Guerrero, R.; Hidalgo, L.; Bevilaqua, D. Evaluation of Biogas Biodesulfurization Using Different Packing Materials. *ChemEngineering* **2019**, *3*, 27.

4. Okoro, O.; Sun, Z. Desulphurisation of Biogas: A Systematic Qualitative and Economic-Based Quantitative Review of Alternative Strategies. *Chemengineering* **2019**, *3*, 76.

5. Le Borgne, S.; Baquerizo, G. Microbial Ecology of Biofiltration Units Used for the Desulfurization of Biogas. *Chemengineering* **2019**, *3*, 72.

 *chemengineering*

Article

Desulfurization of Biogas from a Closed Landfill under Acidic Conditions Deploying an Iron-Redox Biological Process

Antonio Velasco [1], Mariana Franco-Morgado [2], Sergio Revah [3,*], Luis Alberto Arellano-García [4], Matías Manzano-Zavala [3] and Armando González-Sánchez [2,*]

[1]   Departamento de Biotecnología, Universidad Autónoma Metropolitana-Unidad Iztapalapa, Iztapalapa, 09340 Mexico City, Mexico
[2]   Instituto de Ingeniería, Universidad Nacional Autónoma de Mexico, Circuito Escolar, Ciudad Universitaria, 04510 Mexico City, Mexico
[3]   Departamento de Procesos y Tecnología, Universidad Autónoma Metropolitana-Unidad Cuajimalpa, Cuajimalpa, 09340 Mexico City, Mexico
[4]   Cátedras CONACYT-Centro de Investigación y Asistencia en Tecnología y Diseño del Estado de Jalisco, Unidad de Tecnología Ambiental. Av. Normalistas 800, Guadalajara, 44270 Jalisco, Mexico
*   Correspondence: srevah@correo.cua.uam.mx (S.R.); agonzalezs@iingen.unam.mx (A.G.-S.)

Received: 30 April 2019; Accepted: 5 August 2019; Published: 7 August 2019

Abstract: Desulfurization processes play an important role in the use of biogas in the emerging market of renewable energy. In this study, an iron-redox biological process was evaluated at bench scale and pilot scale to remove hydrogen sulfide (H_2S) from biogas. The pilot scale system performance was assessed with real biogas emitted from a closed landfill to determine the desulfurization capacity under outdoor conditions. The system consisted of an Absorption Bubble Column (ABC) and a Biotrickling Filter (BTF) with useful volumes of 3 L and 47 L, respectively. An acidophilic mineral-oxidizing bacterial consortium immobilized in polyurethane foam was utilized to regenerate Fe(III) ion, which in turn accomplished the continuous H_2S removal from inlet biogas. The H_2S removal efficiencies were higher than 99.5% when H_2S inlet concentrations were 120–250 ppmv, yielding a treated biogas with H_2S < 2 ppmv. The ferrous iron oxidation rate ($0.31\ g\cdot L^{-1}\cdot h^{-1}$) attained when the system was operating in natural air convection mode showed that the BTF can operate without pumping air. A brief analysis of the system and the economic aspects are briefly analyzed.

Keywords: biogas; hydrogen sulfide; removal process

1. Introduction

The use of biogas from municipal landfills to obtain energy (electricity generation) is a growing trend worldwide as part of the quest for clean energy alternatives to the traditional fossil fuels [1]. Landfill biogas is produced by the anaerobic digestion of organic wastes, and its composition depends on the type and age of digested organic matter. Typically, landfill biogas is composed of methane (CH_4) 50% v, carbon dioxide (CO_2) 45% v, alkanes/alkenes (C_7H_8–$C_{16}H_{34}$) 0.1–85.3 $mg\cdot m^{-3}$, chlorides (CCl_4–C_2HCl_3) 0.14–4.52 $mg\cdot m^{-3}$, mercury compounds (CH_3Hg–$(CH_3)2Hg$) 1–91 $\mu g\ m^{-3}$, siloxanes 1–17 $mg\cdot m^{-3}$, volatile organic compounds (VOCs) (benzenes, isopropyl benzene, halogenated compounds) 5–85 $mg\cdot m^{-3}$ and hydrogen sulfide 0.005–2% v [2]. The H_2S content depends on the composition and age of the waste disposed in the landfill besides the protein content in organic waste [3,4].

Hydrogen sulfide must be removed from biogas due to technical problems related to corrosion in pipes, pumps, engines, gas storage tanks and electric power plants, as well as the fact of it being a potential pollutant when it is combusted, producing sulfur dioxide (SO_2) [3]. This gas is further oxidized, which promotes acid rain containing sulfuric acid (H_2SO_4) [5]. Additionally, it causes a bad odor at very low concentrations due to its low odor threshold (1 ppbv) [6].

The final use of biogas, composition and flow variability, concentration of H_2S, and the absolute quantity of H_2S to be removed define the requirements of the desulfurization technique to be deployed [4]. To remove the H_2S content in a biogas stream, there are several physicochemical technologies with good removal efficiencies (> 99%) [3]. LO-CAT® technology is an example of a physicochemical technology that has been applied in more than 120 plants around the world [7]. The removal mechanism is based on a series of chemical reactions of H_2S with iron chelating agents under slightly alkaline conditions [8,9]. The products, after the chemical H_2S removal, are elemental sulfur and ferrous ion, the former being recovered by sedimentation, while ferrous ion is continuously oxidized into ferric ion using an inlet air stream [7]. Moreover, under acid conditions (pH < 2) the chemical H_2S reactions with ferric ions can be carried out without chelating agents because iron (both Fe(II) and Fe(III)) remain soluble without sulfide iron precipitation; however, under acidic conditions, the oxidation rate of ferrous ion by molecular oxygen is slow [7]. Certain bacteria, such as *Acidithiobacillus ferrooxidans*, play an important role in increasing the rate of ferrous iron oxidation into ferric iron. Meruane and Vargas [10] showed that at a low pH (pH < 5), the rate of bacterial oxidation of ferrous iron is about 10^4 times larger than the corresponding rate of chemical oxidation. This result indicates that acidophilic iron-oxidizing bacteria, such as *A. ferrooxidans*, are a promising microorganism for usage in desulfurization processes, regenerating Fe(III) biologically [7]. In combination with physicochemical oxidation, *A. ferrooxidans* can act as catalyst of reoxidation of ferrous ions to achieve the removal of H_2S from biogas with lower operational and environmental costs compared with a sole physicochemical technology [9]. Nowadays, biological desulfurization treatments have gained attention due to the achieved removal efficiencies (> 99%) and are competitive with physicochemical methods. Some documented examples of biodesulfurization processes, including biological ferric ion regeneration, are biofilters, biotrickling filters, Biogas Cleaner®, Biopuric®, DMT filter®, LO-CAT® and SulFerox® among others [4,7,11–13]. However, challenges remain in the scaling up of these technologies in terms of the consumption of chelated iron, pH control, and overall economic balance of the process [14].

The aim of this work was to present the experimental performance of an on-site chemical-biological desulfurization system removing H_2S from biogas generated at a closed landfill. The effects of no pH control and no forced convection of air on the iron oxidation rates and H_2S removal were evaluated.

2. Materials and Methods

2.1. Microorganisms

An enriched acidophilic mineral-oxidizing bacterial consortium (AMOB), obtained from the sediments and soil of an acid mine drainage in Taxco Guerrero Mexico, was used as inoculum for the biological oxidation of the ferric ion. The AMOB was grown in medium 9K [15] containing ($g \cdot L^{-1}$): 3.0 $(NH_4)_2 SO_4$, 0.5 $MgSO_4 \cdot 7H_2O$, 0.5 K_2HPO_4, 0.1 KCl, 0.01 $Ca(NO_3)_2$ and 44.8 $FeSO_4 \cdot 7H_2O$ (corresponding to 9.9 g Fe(II) L^{-1}); the pH was adjusted to 1.6 with H_2SO_4.

2.2. Prototype Experimental System

The prototype system was previously tested in lab conditions, feeding controlled H_2S concentrations in defined air flow rates, which were made by mixing fresh air with a controlled flow of pure H_2S. Further details can be found elsewhere [9].

Figure 1 shows the prototype system installed in the closed landfill "Prados de la Montaña" in the western part of Mexico City. The landfill was closed in 1992; however, it continues to produce biogas, and further details can be found elsewhere [16]. The prototype system was connected to a venting-outlet of the landfill through a peristatic pump that supplied the sour gas at a flow of

960 L·d^{-1}. The prototype experimental system called Hybrid System at Pilot Scale (HSPS) consisted of two columns: an absorption bubble column (ABC) and a biotrickling filter (BTF) with useful volumes of 3 L and 47 L, respectively, and interconnected by a recycled aqueous stream. The 960 L·d^{-1} of sour biogas were fed at the bottom of the ABC co-currently with a 777 L·d^{-1} stream of 9K medium coming from the bottom of the BTF. The desulfurized biogas stream obtained from the top of the ABC was captured for a posterior composition analysis. The BTF was packed with polyurethane foam (EDT, Germany) with a specific area of 600 m^2·m^{-3}, a density of 35 kg·m^{-3} and a porosity of 0.97. The BTF was inoculated with the aforementioned AMOB. To keep aerobic conditions in the BTF, either a forced or a natural convective flow of air was implemented by pumping air at a flow of 82,000 L·d^{-1} to the BTF or just by keeping two air vents at extreme opposed sides of the BTF open, respectively. The forced and natural convective airflow tests allowed for the evaluation of the re oxidizing rates of ferrous ions with a minimum input of energy for aeration. The 9K medium was trickled from the top of the BTF with a flow of 3740 L·d^{-1}. The pH was maintained at 1.2 without an automatic control, and the temperature oscillated between 5 and 30 °C due to the outdoor conditions prevailing in Mexico City. The water evaporation was compensated daily with fresh water, while the 9K medium was renewed every 3 months. The HSPS was operated continuously for around 7 months.

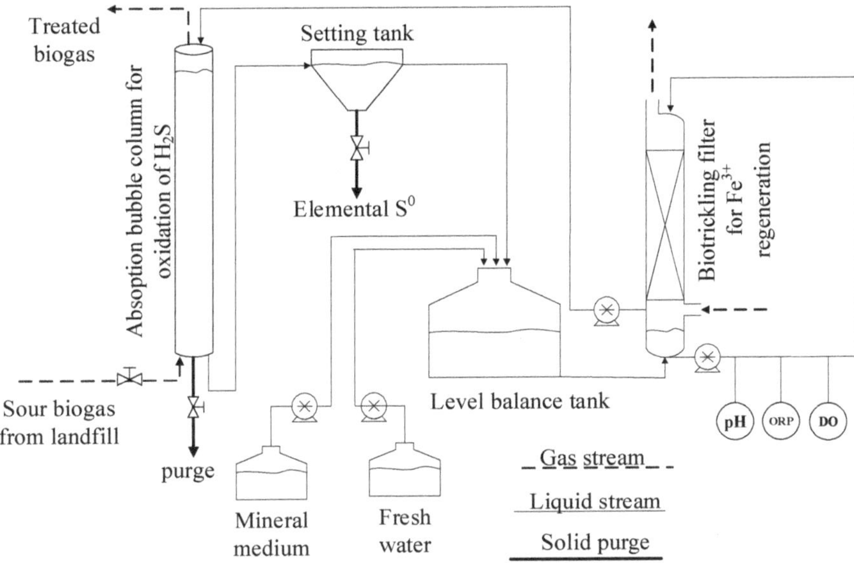

Figure 1. Desulfurization Hybrid System at Pilot Scale (HSPS) installed on a landfill cover.

The Fe(III) ion regeneration rate in the BTF was evaluated under a batch operation, with an initial Fe(II) concentration of around 4.5 L·d^{-1} under the continuous recycling of the 9K aqueous medium at the conditions described above.

The predominance zones diagrams for the stable iron and sulfur species with water under the experimental conditions were calculated with the software HSC Chemistry® Version 4.1 (Outokumpu Research Oy, Pori, Finland). The software computes the predominance zones in the pH vs. Oxidation-Reduction Potential (ORP) graph (also called Pourbaix diagram) at equilibrium. The software inputs are the total molal sulfur and iron concentrations in the aqueous phase contained in the HSPS, as well as the system conditions (temperature and pressure).

2.3. Analytical Methods

During the operation of the desulfurization system, the gas phase H_2S concentrations were continuously measured using an Odalog sensor with a range of 1–1000 ppm (App-Tek, distributed by Detection Instruments, Phoenix, AZ), which included a temperature sensor. In the aqueous phase, the total iron concentration in the 9K recycling medium was measured by titration with potassium dichromate and barium diphenylamine-sulfonate according to the method reported by Vogel [17]. Samples were collected from the bottom of the BTF and in the ABC. The ORP was measured with a polished platinum probe, using an Ag/AgCl electrode as a reference (EW-27018-40, Cole Parmer, Vernon Hills, IL, USA). The dissolved oxygen (DO) was monitored through a polarographic probe (Hanna Instruments, Woonsocket, RI, USA), and both ORP and DO were recorded online by means of a personal computer.

3. Results and Discussion

3.1. Oxidation of Ferrous Iron in the BTF

Figure 2A shows the depletion of Fe(II) in the culture medium under continuous forced air supply to the BTF. In this experiment, the Fe(II) oxidation rate was around 8.16 $g \cdot L^{-1} \cdot d^{-1}$, which, compared with the value of 0.19 $g \cdot L^{-1} \cdot h^{-1}$ reported by Daoud and Karamanev [18], it shows that ferrous iron-oxidizing bacteria consortia used in our study have an adequate response to the reactor fixed conditions (pH, nutrients, airstream, flow recirculation, etc.). Moreover, Figure 2B shows that Fe(II) oxidation in the BTF was effective both under air forced convection (initial 250 min of experiment) and under the natural convection mode (final 130 min). The ferrous iron oxidation rate was calculated as 7.44 $g \cdot L^{-1} \cdot d^{-1}$ in the natural convection mode, being similar to the rate obtained with the forced air convection (9.12 $g \cdot L^{-1} \cdot d^{-1}$). These results suggested that the BTF can operate under a natural convection mode to accomplish the biological oxidation of Fe(II). The dissolved oxygen concentration in the trickling liquid remained constant around 0.0065 $g \cdot L^{-1}$ for both assays, confirming that for this BTF the oxygen mass transfer did not limit the Fe(II) biological oxidation.

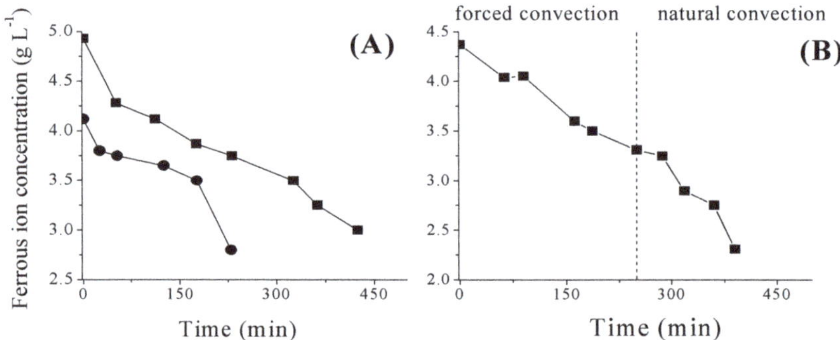

Figure 2. Biological oxidation of Fe(II) to Fe(III) in (**A**) the forced convection mode (black squares: assay 1, black circles: assay 2) and (**B**) the forced/natural convection mode (black squares: mixed assay).

3.2. Removal of H_2S in the Prototype Hybrid System under Lab Conditions

During the lab assays, Figure 3A,B show the results of the H_2S elimination capacity at inlet concentrations of 500 and 1000 ppmv respectively fed to the prototype hybrid system operated at the natural air convection mode for 120 h.

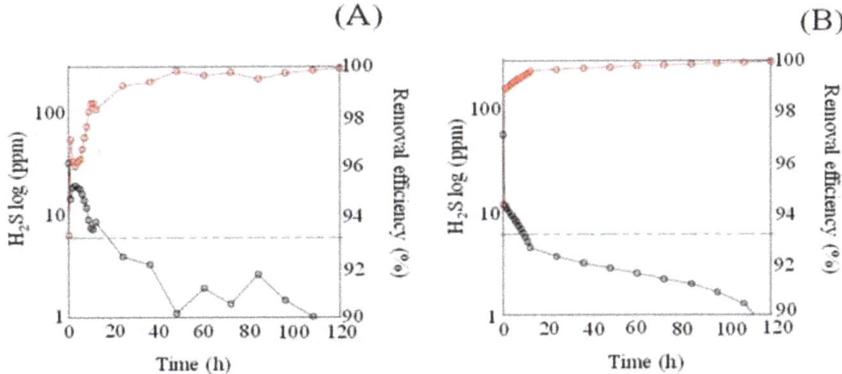

Figure 3. The removal efficiency of H_2S at inlet concentrations of (**A**) 500 ppmv and (**B**) 1000 ppmv. Red circles: % removal efficiency; black circles: outlet H_2S concentration.

The results show that the outlet gaseous H_2S concentrations decrease through time and that this is sustained due to the continuous chemical reaction of H_2S with Fe(III) and the subsequent biological oxidation of Fe(II) into Fe(III). After the first 20 h, the average elimination capacity of H_2S was 99% for both 500 and 1000 ppmv of H_2S. Ho et al. [19] reported H_2S removal efficiencies of around 82% under similar gaseous residence times (4 min), from inlet H_2S concentrations of 1500 ppmv using ferric iron concentration between 9 and 11 $g·L^{-1}$. These authors report that *A. ferrooxidans* CP9 registered a high iron tolerance, up to 20 $g·L^{-1}$. In our study, the pH between 1.6 to 2.0 at all times indicated conditions where the ferric ion precipitation is minimized [7], allowing iron ions to be continuously recycled to react with H_2S in the ABC and with O_2 in the BTF.

In our study, an average Fe(II) oxidation of 7.44 $g·L^{-1}·d^{-1}$ was estimated, showing that the biological system provided a sufficient ferric iron regeneration for the stable and efficient H_2S elimination from landfill gas. In comparison, recent reports show Fe(II) oxidation rates of 2.0 $g·L^{-1}·d^{-1}$ and 7.2 $g·L^{-1}·d^{-1}$ in batch and 161 $g·L^{-1}·d^{-1}$ in continuous systems [11–13]. This data was important for establishing the optimal operating parameters to scale up the hybrid system. Regarding the distribution of the iron species, a Pourbaix diagram of the system sulfur-iron-water was computed. This diagram considered the total iron concentration in the aqueous phase of the prototype system (4.5 $g·L^{-1}$, 8.00×10^{-2} M) and the maximum dissolved sulfur concentration (2.17×10^{-5} M), estimated from both the H_2S aqueous solubility and average concentration close to 200 ppmv in the closed landfill vent, at an average temperature of 30 °C. The predominance zones diagram is shown in Figure 4. It can be seen that, theoretically, the Fe(III) ion is stable at a very narrow zone delimited by very specific conditions of ORP above +0.55 V and a pH below 0.5. However, when H_2S is absorbed in an aqueous solution at an acidic pH below 4.5, it reacts with $2FeOH^{2+}$, producing elemental sulfur (S^0), $2Fe^{2+}$ and water, as was recently described [20–22]. In our study, it is likely that this mechanism predominates under the conditions of our experimentation. This metastable zone is marked with a red circle on Figure 4. In our process, uncontrolled pH showed an average value of 1.8, while the mean ORP value was close to +0.5 V. Other studies reported that the optimum absorption rate of H_2S in ferric iron solutions occurs at pH 2.2 and that the absorption rate at pH 1.6 is expected to be approximately 50% lower [23]. In spite of this, we demonstrated satisfactory absorption H_2S rates with no need of extra equipment and reactants for the pH control.

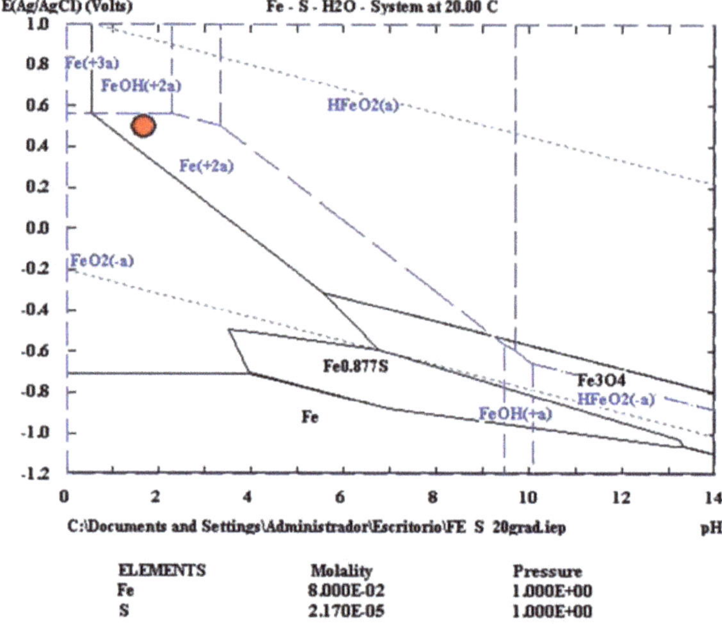

ELEMENTS	Molality	Pressure
Fe	8.000E-02	1.000E+00
S	2.170E-05	1.000E+00

Figure 4. Predominance diagram zones at a total iron concentration of 8.00×10^{-2} M, total sulfur concentration of 2.17×10^{-5} M, and total pressure of 1 atm.

3.3. Performance of the Hybrid System at Pilot Scale (HSPS) in the Landfill

Figure 5 shows the inlet H_2S concentration and removal efficiency obtained during 50 d of operation of the HSPS installed in the closed landfill, where biogas contained a H_2S concentration between 120 to 250 ppmv. The results showed that the HSPS achieved an H_2S removal of up to 99.5% during the 50 d of sour biogas feeding. This removal efficiency guaranteed that H_2S in the treated biogas was below 2 ppmv. The pH and ORP values of the solution remained relatively constant around 2.0 and +0.5 V, respectively, as it was measured continuously during one week (see Figure 6B). In addition, no negative effect was registered on the biological process due to the exposition to outdoor conditions prevailing in the landfill, i.e., a diurnal temperature variation of 15 and 30 °C during the period of evaluation. The high H_2S removal efficiencies registered by the HSPS in the landfill were expected, as the concentration of H_2S in the sour biogas coming from the landfill and assayed during the lab evaluation stage never exceeded 500 ppmv, as shown in Figure 2A. De La Rosa et al. [24] mentioned that the biogas produced in this site contained small amounts of N_2, O_2, NH_3, H_2, CO, and traces of toxic substances (VOCs, mercaptans, gaseous mercury). In this respect, Zhang et al. [25] indicated that iron-oxidizing bacteria such as *A. ferrooxidans* can tolerate diverse organic compounds and metal ions within a certain concentration, which permitted survival and growth in extreme environments such as metal mines, coal mines, and sewage treatment plants. The good performance of the pilot plant in the landfill, registered during the on-site evaluation, showed the robustness of the combination of chemical and biological processes under the operational conditions here described. It is important to mention that the HSPS required minimum operational services, i.e., no aeration (natural convection) nor any pH or temperature control.

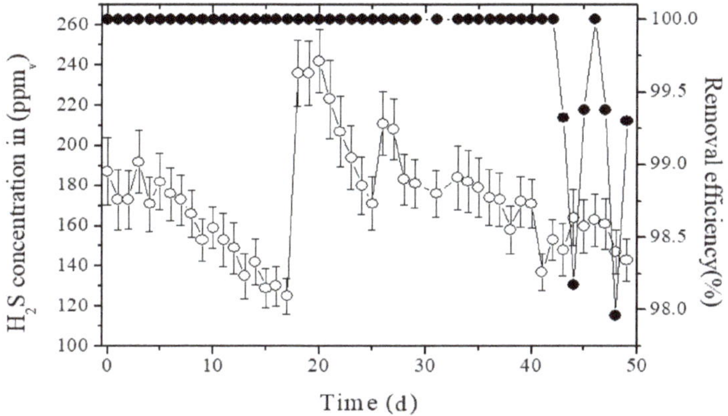

Figure 5. Performance of the HSPS in the landfill. H$_2$S inlet (○) and removal efficiency (●).

The inlet concentration of H$_2$S and the profile of the temperature of the biogas fed to the HSPS during the first week (Figure 6) showed that periodic variation of both parameters occurred through time. In addition, Figure 6 shows a decrease in the H$_2$S concentration in the biogas coming from the landfill, especially during the rainy season, probably due to the increment of moisture inside the landfill, which may cause a disequilibrium of the H$_2$S partition between gas and liquid in the landfill. However, these environmental factors do not affect the efficiency of the hybrid system, in particular the temperature variation, which can cause a negative effect on the biological activity.

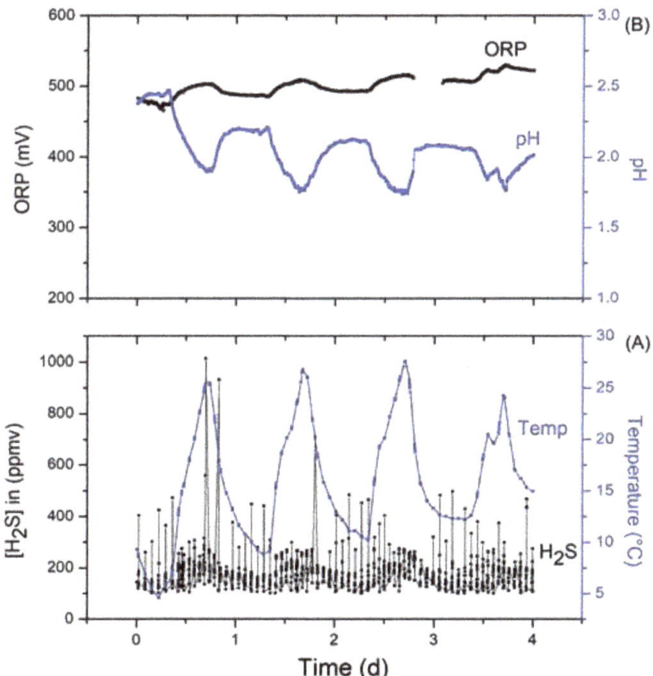

Figure 6. (**A**) The inlet concentration of H$_2$S and temperature profile; (**B**) the pH and Oxidation-Reduction Potential (ORP) profile of the aqueous solution in the HSPS, during the first week of November 2011.

Sulfate concentrations of around 10 g·L^{-1} were detected after one month of continuous operation of the HSPS on the landfill, probably as a result of the ability of these bacteria to oxidize sulfide species [6] to obtain energy. We did not recover any elemental sulfur in the settler due to its colloidal property, so the produced elemental sulfur accumulated in the BTF. This characteristic might signify an important advantage for the system in removing an excessive accumulation of elemental sulfur in the packed bed of the BTF, where a temporal interruption of the feeding of sour biogas can promote the consumption of this solid sulfur.

The HSPS operation under the aforementioned conditions needed 0.472 kW for its operation, corresponding mainly to the pumps consumption. However, the energy generation by burning 960 L·d^{-1} of biogas was calculated at 0.181 kW (considering a calorific value of 4.5 kWh per m^3 of treated biogas). This energetic analysis would indicate that under the present operational conditions the implementation of this technology is not energetically feasible using grid electricity to supply the system. Moreover, this analysis indicates that this biogas purification process would start to be energetically feasible when treating biogas flows greater than 2400 L·d^{-1}. In this scenario, the implementation of alternative energy sources such as solar and wind would allow to make this energetic balance positive. Besides, the desulfurization of biogas remains important, keeping in mind that reducing methane emissions to the atmosphere might be accomplished by the combustion of desulfurized biogas, which can further aid the combustion equipment lifespan while also producing clean energy.

4. Conclusions

The prototype deployed for the desulfurization of biogas emitted from a closed landfill at the pilot scale demonstrated comparable desulfurization capacities with other biological desulfurization processes; with the added advantage of requiring a minimal energy input. A metastable operation zone between the 2FeOH^{2+} and Fe^{2+} was identified, considering the pH and ORP reached during the continuous desulfurization of sour biogas emitted from the closed landfill Prados de la Montaña in Mexico City. The capacity of treatment and then its energetic feasibility could be improved by finding the limits of treatment without increasing the operational energy consumption.

Author Contributions: Conceptualization, S.R., A.G.-S.; methodology, A.V., M.F.-M., L.A.A.-G., M.M.-Z., S.R., A.G.-S.; investigation, A.V., M.F.-M., L.A.A.-G., M.M.-Z.; resources, S.R., A.G.-S.; A.V., L.A.A.-G.; writing, A.V., L.A.A.-G., A.G.-S.; writing—review and editing, A.G.-S., S.R.; supervision, A.G.-S.; project administration, S.R.; funding acquisition, S.R., A.G.-S.

Funding: This research was funded by the former Science and Technology Institute of Mexico City (grant # ICyTDF 285/2009)

Acknowledgments: César Sánchez for his technical support.

Conflicts of Interest: The authors declare no conflict of interest.

References

1. Kigozi, R.; Muzenda, E.; Aboyade, A.O. Biogas technology: Current trends, opportunities and challanges. In Proceedings of the 6th International Conference GREEN Technology, Renewable Energy & Environmental Engineering, Cape Town, South Africa, 27–28 November 2014; pp. 311–317.
2. Kuhn, J.N.; Elwell, A.C.; Elsayed, N.H.; Joseph, B. Requirements, techniques, and costs for contaminant removal from landfill gas. *Waste Manag.* **2017**, *63*, 246–256. [CrossRef] [PubMed]
3. Awe, O.W.; Zhao, Y.; Nzihou, A.; Minh, D.P.; Lyczko, N. A Review of Biogas Utilisation, Purification and Upgrading Technologies. *Waste Biomass Valoriz.* **2017**, *8*, 267–283. [CrossRef]
4. Bailon, A.L.; Hinge, J. *Biogas Upgrading Evaluation of Methods for H$_2$S Removal*; Danish Technological Institute: Taastrup, Denmark, 2014.
5. Latosov, E.; Loorits, M.; Maaten, B.; Volkova, A.; Soosaar, S. Corrosive effects of H$_2$S and NH$_3$ on natural gas piping systems manufactured of carbon steel. *Energy Procedia* **2017**, *128*, 316–323. [CrossRef]
6. González-Sánchez, A.; Revah, S.; Deshusses, M.A. Alkaline biofiltration of H$_2$S odors. *Sci. Technol.* **2008**, *42*, 7398–7404. [CrossRef] [PubMed]

7. De Angelis, A.; Bellussi, G.; Pollesel, P.; Perego, C. New method for H_2S removal in acid solutions. *ChemSusChem* **2010**, *3*, 829–833. [CrossRef] [PubMed]

8. Devinny, J.S.; Deshusses, M.A.; Webster, T.S. *Biofiltration for Air Pollution Control*, 1st ed.; Lewis Publishers: Boca Raton, FL, USA, 2017.

9. Velasco, A.; Morgan-Sagastume, J.M.; González-Sánchez, A. Evaluation of a hybrid physicochemical/biological technology to remove toxic H_2S from air with elemental sulfur recovery. *Chemosphere* **2019**, *222*, 732–741. [CrossRef] [PubMed]

10. Meruane, G.; Vargas, T. Bacterial oxidation of ferrous iron by Acidithiobacillus. *Hydrometallurgy* **2003**, *71*, 149–158. [CrossRef]

11. Alemzadeh, I.; Kahrizi, E.; Vossoughi, M. Bio-oxidation of ferrous ions by *Acidithiooobacillus ferrooxidans* in a monolithic bioreactor. *J. Chem. Technol. Biotechnol.* **2009**, *8*, 504–510. [CrossRef]

12. Almenglo, F.; Ramírez, M.; Gómez, J.M.; Cantero, D.; Revah, S.; González-Sánchez, A. Effect of VOCs and methane in the biological oxidation of the ferrous ion by an acidophilic consortium. *Environ. Technol.* **2012**, *335*, 531–537. [CrossRef]

13. Lin, W.-C.; Chen, Y.-P.; Tseng, C.-P. Pilot-scale chemical–biological system for efficient H_2S removal from biogas. *Bioresour. Technol.* **2013**, *135*, 283–291. [CrossRef]

14. Abatzoglou, N.; Boivin, S. A review of biogas purification processes. *Biofuels Bioprod. Bioref.* **2009**, *3*, 42–71. [CrossRef]

15. Silverman, M.P.; Lundgren, D.G. Studies on the chemoautotrophic iron bacterium Ferrobacillus ferrooxidans: I. An improved medium and a harvesting procedure for securing high cell yields. *J. Bacteriol.* **1959**, *77*, 642. [PubMed]

16. Villanueva-Estrada, R.E.; Rocha-Miller, R.; Arvizu-Fernández, J.L.; Castro González, A. Energy production from biogas in a closed landfill: A case study of Prados de la Montaña, Mexico City. *Sustain. Energy Technol. Assess.* **2019**, *31*, 236–244. [CrossRef]

17. Vogel, A.I. *A Text-Book of Quantitative Inorganic Analysis-Theory and Practice*; Longmans, Green and Co.: London, UK; New York, NY, USA; Toronto, ON, Canada, 2013.

18. Daoud, J.; Karamanev, D. Formation of jarosite during Fe^{2+} oxidation by *Acidithiobacillus ferrooxidans*. *Miner. Eng.* **2006**, *19*, 960–967. [CrossRef]

19. Ho, K.L.; Lin, W.C.; Chung, Y.C.; Chen, Y.P.; Tseng, C.P. Elimination of high concentration hydrogen sulfide and biogas purification by chemical-biological process. *Chemosphere* **2013**, *92*, 1396–1401. [CrossRef] [PubMed]

20. Asai, S.; Yasuhiro, K.; Tadahiro, Y. Kinetics of absorption of hydrogen sulfide into aqueous Fe(III) solutions. *AIChE J.* **1990**, *36*, 1331–1338. [CrossRef]

21. Demmink, J.F.; Beenackers, A.A.C.M. Gas desulfurization with ferric chelates of EDTA and HEDTA. *Ind. Eng. Chem. Res.* **1998**, *37*, 1444–1453. [CrossRef]

22. Wubs, J.H.; Beenackers, A.A.C.M. Kinetics of the Oxidation of Ferrous Chelates of EDTA and HEDTA in Aqueous Solution. *Ind. Eng. Chem. Res.* **1993**, *11*, 2580–2594. [CrossRef]

23. Asai, S.; Nakamura, H.; Aikawa, H. Absorption of hydrogen sulfide into aqueous ferric chloride solutions. *J. Chem. Eng. Jpn.* **1997**, *30*, 500–506. [CrossRef]

24. De La Rosa, D.A.; Velasco, A.; Rosas, A.; Volke-Sepúlveda, T. Total gaseous mercury and volatile organic compounds measurements at five municipal solid waste disposal sites surrounding the Mexico City Metropolitan Area. *Atmos. Environ.* **2006**, *40*, 2079–2088. [CrossRef]

25. Zhang, S.; Yan, L.; Xing, W.; Chen, P.; Zhang, Y.; Wang, W. *Acidithiobacillus ferrooxidans* and its potential application. *Extremophiles* **2018**, *22*, 563–579. [CrossRef]

 *chemengineering*

Article

Application of Response Surface Methodology for H$_2$S Removal from Biogas by a Pilot Anoxic Biotrickling Filter

Fernando Almenglo *, **Martín Ramírez** * and **Domingo Cantero**

Department of Chemical Engineering and Food Technologies, Wine and Agrifood Research Institute (IVAGRO), Faculty of Sciences, University of Cádiz, 11510 Puerto Real (Cádiz), Spain
* Correspondence: fernando.almenglo@uca.es (F.A.); martin.ramirez@uca.es (M.R.); Tel.: +34-956016286 (M.R.)

Received: 13 May 2019; Accepted: 11 July 2019; Published: 13 July 2019

Abstract: In this study, a pilot biotrickling filter (BTF) was installed in a wastewater treatment plant to treat real biogas. The biogas flow rate was between 1 and 5 m^3·h^{-1} with an H$_2$S inlet load (*IL*) between 35.1 and 172.4 gS·m^{-3}·h^{-1}. The effects of the biogas flow rate, trickling liquid velocity (*TLV*) and nitrate concentration on the outlet H$_2$S concentration and elimination capacity (*EC*) were studied using a full factorial design (3^3). Moreover, the results were adjusted using Ottengraf's model. The most influential factors in the empirical model were the *TLV* and H$_2$S *IL*, whereas the nitrate concentration had less influence. The statistical results showed high predictability and good correlation between models and the experimental results. The R-squared was 95.77% and 99.63% for the '*C model*' and the '*EC model*', respectively. The models allowed the maximum H$_2$S *IL* (between 66.72 and 119.75 gS·m^{-3}·h^{-1}) to be determined for biogas use in a combustion engine (inlet H$_2$S concentration between 72 and 359 ppm$_V$). The '*C model*' was more sensitive to *TLV* (–0.1579 (gS·m^{-3})/(m·h^{-1})) in the same way the '*EC model*' was also more sensitive to *TLV* (4.3303 (gS·m^{-3})/(m·h^{-1})). The results were successfully fitted to Ottengraf's model with a first-order kinetic limitation (R-squared above 0.92).

Keywords: hydrogen sulfide; anoxic biotrickling filter; biogas; Ottengraf's model; open polyurethane foam; response surface methodology

1. Introduction

Biogas, due to methane high combustion enthalpy, can be considered as an important renewable energy source. Nowadays, international laws such as the Directive of the European Parliament 2009/28/EC (April 23, 2009) recognize biogas as a source of vital energy that can reduce the European Union's energy dependence. The aim of this directive is to increase consumption of renewable energy by 2020 by at least 20%. The longer-term goal is to achieve net-zero greenhouse emissions by 2050 and it will be necessary to increase investment in clean and energy-efficient technologies by 2.8% of Gross Domestic Product (GDP) (or around € 520–575 billion annually) [1]. Hydrogen sulfide (H$_2$S) is one of the biggest pollutants in biogas. Therefore, it is necessary to reduce the generation of H$_2$S in the digester and/or reduce its concentration for most uses of biogas. Apart from its harmful effects on health, the presence of H$_2$S in biogas is not desirable because it is a corrosive gas. Desulfurization of biogas can be carried out by physical-chemical or biological processes. Physical-chemical processes have been commonly used but biological ones have proven to be a good competitor from economic and environmental points of view [2].

In biological processes the most widely used and studied microorganisms belong to the group of bacterial chemotrophic species, which use reduced sulfur compounds as an energy source and use oxygen (aerobic) or nitrate/nitrite (anoxic) as electron acceptors [3,4]. The biodesulfurization of biogas involves the use of technologies that facilitate appropriate contact between the gas and the liquid.

There are several advantages in the use of anoxic biotrickling filters (BTFs) over aerobic ones, and these include reducing the risk of explosion, no dilution of biogas, and a lower limitation in the transfer of matter for nitrate when compared to the necessary oxygen absorption [5,6]. In contrast, the cost and availability of large amounts of nitrate can be limiting for the application of anoxic systems. Although ammonia-rich wastewater could be nitrified and used, Zeng et al. [3] used a biogas digestion slurry after nitrification to feed a BTF to achieve stable operation.

The H_2S inlet load (*IL*) in BTFs is an important parameter in the design of this type of equipment. The *IL* and the elimination capacity (*EC*) describe the performance of the BTFs with respect to the removal efficiency (*RE*) of the contaminant and allow the design of the system, depending on the biogas flow rate that needs to be treated and the inlet H_2S concentration. Values for a critical *EC* of between 100 and 130 $gS \cdot m^{-3} \cdot h^{-1}$ and a maximum *EC* between 140 and 280 $gS \cdot m^{-3} \cdot h^{-1}$ have been reported for both aerobic and anoxic biotrickling filters [5,7].

A model can be applied to relate input variables (pollutant inlet concentration, gas flow, electron acceptor concentration, etc.) and design variables (specific surface area of the support, equipment dimensioning, etc.) with the outputs (concentration of the pollutant at the outlet, production of biological reaction products, consumption of reagents, etc.). Empirical models (black box models) are characterized by a high predictive power, but their parameters lack physical significance [8]. They are based on statistically significant relationships between the input and output variables. Stationary-state models have been used to describe biofilters since the early 1980s [9–11]. Anoxic BTFs have been described by empirical [12,13] and dynamic [14,15] models. For instance, Soreanu et al. [13] proposed a second-order empirical model using a central composite design (CCD) and the biogas flow rate and the H_2S concentration as input variables, with the H_2S *RE* obtained as a response variable. Almenglo et al. [15] developed a model that considered the most relevant phenomena such as advection, absorption, diffusion and biodegration. Dynamic models provide a better understanding of the process, but their complexity means that they are seldom used for BTFs.

The aim of the work described here was to study the effects of gas (F_G) and liquid (F_L) flow rates and nitrate concentration ([N-NO$_3^-$]) along the packed bed on the outlet H_2S concentration and the *EC*. Two empirical models were proposed to describe the outlet H_2S concentration and the *EC*. Moreover, the H_2S concentrations were measured along the bed and fitted using Ottengraf's model [9].

2. Materials and Methods

2.1. Experimental Set-up

An anoxic BTF at pilot scale (Figure 1) was installed in the wastewater treatment plant (WWTP) 'Bahía Gaditana' (San Fernando, Spain) and it was fed with biogas from one of their sludge anaerobic digesters. The internal diameter was 0.5 m and the bed height was 0.85 m. The BTF was packed with open-pore polyurethane foam cubes (800 units, 25 $kg \cdot m^{-3}$, 600 $m^2 \cdot m^{-3}$) (Filtren TM25450, Recticel Iberica, Spain). The recirculation medium pH was kept at 7.4 and the temperature at 30 °C.

The nitrate feeding was done in batch mode and automatized by oxide reduction potential (ORP) (setpoint of −365 mV) [16]: when nitrate was exhausted a fixed liquid volume (25 L) was purged, then a nitrate solution (500 $gNaNO_3 \cdot L^{-1}$) was added to the recirculation medium and finally treated water from the WWTP was added to get a working volume. The nitrate depletion time (NDT) was the time between nitrate feedings, i.e., the time in which microorganisms consumed the nitrate added. NDT was dependent of *IL* and maximum nitrate concentration reached in the nitrate feeding cycle. The volume of the nitrate solution added was modified in concordance to maintain an NDT between 3 and 4 h. This volume ranged in a linear manner between 0.14 and 0.7 L for an H_2S *IL* between 33.3 and 177 $gS \cdot m^{-3} \cdot h^{-1}$. During the nitrate depletion time the H_2S concentrations in the outlet stream and along the bed height (0.2, 0.4, 0.6 and 0.8 m) were measured every 30 min, and samples of the recirculation medium were taken for nitrate, nitrite and sulfate measurement. Further information about the system can be found elsewhere [15,16]. A schematic representation of the experimental set-up is provided in Figure 2.

Figure 1. Photograph of the pilot scale anoxic biotrickling filter.

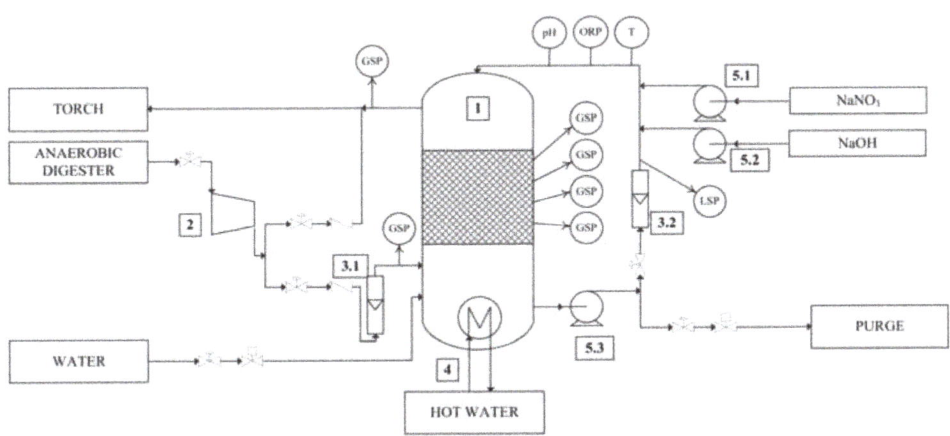

Figure 2. Experimental set-up. GSP—Gas Sampling Port; LSP—Liquid Sampling Port. 1: Biotrickling filter; 2: biogas compressor; 3: rotameters (3.1: biogas and 3.2 liquid); 4: Cryostat Bath; 5: pumps (5.1 $NaNO_3$; 5.2 NaOH; 5.3 recirculation pump).

The H_2S concentration in the biogas stream was measured using a gas chromatograph with a thermal conductivity detector (GC-450, Bruker, Germany) and a specific gas sensor (GAsBadge®Pro, Industrial Scientific, USA) was used for H_2S concentrations below 500 ppm$_V$. Sulfate, nitrite and nitrate were measured by a turbidimetric method (4-500-$SO_4{}^{2-}$ E), a colorimetric method (45000-$NO_2{}^-$ B) and by an ultraviolet method (4500-$NO_3{}^-$ B), respectively [17].

2.2. Experimental Design

A response surface model from a full factorial three-level three-factor design (3^3) was developed including two replicates at the central point. 3^3 design allows us to obtain a second-order polynomial using only three levels [18,19]. The three factors were gas flow rate (F_G), liquid flow rate (F_L) and nitrate concentration. The levels of the factor studied, and the values calculated for H_2S *IL* (Equation (1)), trickling liquid velocity (*TLV*) (Equation (2)) and empty bed residence time (*EBRT*) (Equation (3)) are provided in Table 1.

$$IL = \frac{F_G}{V} \cdot [H_2S]_i \tag{1}$$

$$TLV = \frac{F_L}{A}, \tag{2}$$

$$EBRT = \frac{V}{F_G}, \tag{3}$$

$$EC = IL \cdot RE, \tag{4}$$

$$RE = \frac{[H_2S]_i - [H_2S]_o}{[H_2S]_i}, \tag{5}$$

where *IL* is the inlet load ($gS \cdot m^{-3} \cdot h^{-1}$), *V* is the bed volume (m^3), $[H_2S]_i$ is the inlet H_2S concentration ($gS \cdot m^{-3}$), $[H_2S]_o$ is the outlet H_2S concentration ($gS \cdot m^{-3}$), *RE* is the removal efficiency, *EC* is the elimination capacity, *TLV* is the trickling liquid velocity ($m \cdot h^{-1}$), *EBRT* is the empty bed residence time (s) and *A* is the cross-sectional area of the bed (m^2).

Table 1. Levels of the factor tested in the experimental design (3^3).

Factor	Levels of Factors		
	(−1)	(0)	(+1)
F_G ($m^3 \cdot h^{-1}$)	1	3	5
F_L ($m^3 \cdot h^{-1}$)	1	2	3
[N-$NO_3{}^-$] ($mg \cdot L^{-1}$)	1.4 ± 1.1	35.3 ± 2.4	70.5 ± 10.2
IL [1] ($gS \cdot m^3 \cdot h^{-1}$)	35.1 ± 1.5	109.1 ± 11.7	172.4 ± 3.4
TLV [2] ($m \cdot h^{-1}$)	5.09	10.18	15.27
$EBRT$ [1] (s)	600	200	120

[1] Values calculated for F_G values of 1, 2, and 5 $m^3 \cdot h^{-1}$, respectively, [2] Values calculated for F_L values of 1, 2 and 3 $m^3 \cdot h^{-1}$, respectively.

The experimental results were fitted with two empirical models. The first model, the concentration model ('C *model*'), fitted the outlet H_2S concentration as a response variable. The second model, the elimination capacity model ('EC *model*'), fitted the *EC* as a response variable. In both models the independent variables were *TLV*, the H_2S *IL* and the nitrate concentration (factors from Table 1). Instead of F_G, H_2S *IL* was chosen because the *IL* included the effect of the inlet H_2S concentration (Equation (1)). In addition, *TLV* was used rather than F_L because it allows a comparison with other BTFs. In both cases a second-order polynomial model was used to predict the outlet H_2S concentration and the *EC* values. The data were analyzed using Statgraphics® Centurion XVIII (v.18.1.10).

2.3. Ottengraf's Model

Ottengraf's model [9,20] describes the concentration of pollutants in biofilters for steady-state processes. The analytical solution of the model was obtained in three ideal situations:

1. There is no diffusion limitation and the biofilm is fully active, and hence the conversion rate is controlled by a zero-order reaction rate. The solution is described by Equation (6). K_0 is a pseudo zero-order rate ($g \cdot m^{-3} \cdot s^{-1}$) and is proportional to the zero-order reaction rate constant (k_0, Equation (7)).

2. There is diffusion limitation and therefore the mass transfer rate to the biofilm is insufficient compared to biological substrate utilization rate. The solution is described by Equation (8).

3. There is no diffusion limitation and the biofilm is fully active, and hence the conversion rate is controlled by a first-order reaction rate. The solution is described by Equation (6). K_1 is a pseudo first-order rate constant (s^{-1}) (Equation (10)) and is proportional to the first-order rate constant (k_1, Equation (10)).

$$\frac{[H_2S]}{[H_2S]_0} = 1 - \frac{K_0 \cdot H}{[H_2S]_0 \cdot U_G}, \tag{6}$$

$$\text{With, } K_0 = \delta \cdot A_S \cdot k_0, \tag{7}$$

$$\frac{[H_2S]}{[H_2S]_0} = \left[1 - \frac{H}{U_G} \sqrt{\frac{A_S \cdot K_0 \cdot D'}{2 \cdot [H_2S]_0 \cdot m \cdot \delta}}\right]^2, \tag{8}$$

$$\frac{[H_2S]}{[H_2S]_0} = exp\left(-\frac{H \cdot K_1}{m \cdot U_G}\right), \tag{9}$$

$$\text{With, } K_1 = A_S \cdot D'/\delta \cdot \phi \cdot \tanh\phi, \text{ and } \phi = \delta \sqrt{k_1/D'} \tag{10}$$

where, H is the height of the tower (m), $[H_2S]_0$ is the inlet concentration ($g \cdot m^{-3}$), U_G is the superficial gas velocity ($m \cdot s^{-1}$), A_S is the specific area ($m^2 \cdot m^{-3}$), δ is the biofilm thickness (m), m is the distribution coefficient at equilibrium ($m = G_G/C_L$) and D' is the effective diffusion coefficient ($m^2 \cdot s^{-1}$).

3. Results and Discussion

3.1. Empirical Model

The sulfate and nitrite concentrations were almost constant during the experiments; the sulfate concentration was 8.9 ± 1.6 gS-SO$_4 \cdot$L^{-1} and nitrite concentration was between 0.1 and 10 mgN-NO$_2^- \cdot$L^{-1}. The H$_2$S *RE* obtained during the experimentation carried out to obtain the empirical model is shown in Figure 3.

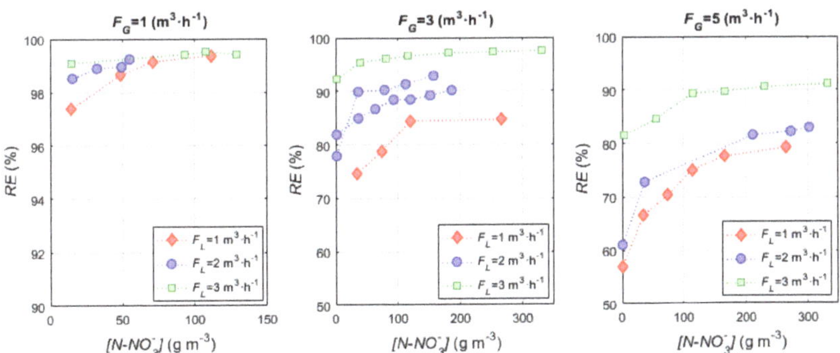

Figure 3. Removal efficiency (*RE*) versus nitrate concentrations at different biogas (F_G) and recirculation (F_L) flow rates.

As expected, a high H$_2$S *RE* was found at low F_G (lower H$_2$S *IL*). Therefore, the best results were for an F_G of 1 m$^3 \cdot$h^{-1}, where the H$_2$S *IL* was 35.1 ± 1.5 gS$\cdot$m$^{-3} \cdot$h^{-1} and the *RE* between 97.3 and 99.5%.

Under these conditions, the effects of the nitrate concentration and *TLV* were negligible. However, at higher biogas flow rates the decrease in the nitrate concentration led to a lower H_2S *RE*. In anoxic biofiltration nitrate (or nitrite) is the electron acceptor, so its concentration must be a significant factor on the BTF performance, and this is even more important considering that the use of a nitrate feed controlled by ORP [16] leads to a decrease in the nitrate concentration until depletion.

The *RE* versus the *TLV* values for the three F_G (*EBRT* of 600, 200 and 120 s) are listed in Table 2. When the nitrate concentration was not limiting, the *TLV* effect on H_2S *RE* was only notable for an F_G equal to or greater than 3 $m^3 \cdot h^{-1}$ (i.e. for H_2S *IL* of 109.1 ± 11.7 and 172.4 ± 3.4 $gS \cdot m^{-3} \cdot h^{-1}$). Thus, for an F_G of 3 $m^3 \cdot h^{-1}$, the H_2S was between 84.7 and 97.6% and for 5 $m^3 \cdot h^{-1}$ the range was between 79.3 and 92.1%. The improvement observed could be explained by various effects: a higher wetted area [21], an increase in the hold-up liquid (6.4, 8.5 and 10.6 L for 5.1, 10.2 and 15.3 $m \cdot h^{-1}$) and a higher area in contact with the flowing liquid, as proposed by Almenglo et al. [15].

Table 2. Removal efficiencies at different trickling liquid velocity (*TLV*s) and empty bed residence time (*EBRT*).

TLV ($m \cdot h^{-1}$)	*EBRT* (s) [1]		
	600	200	120
5.09	99.39	84.73	79.31
10.18	99.24	93.08	83.06
15.27	99.53	97.65	92.13

[1] The H_2S *IL*s were 35.1±1.5, 109.1±11.7 and 172.4±3.4 $gS \cdot m^{-3} \cdot h^{-1}$, for 600, 200 and 120 s, respectively.

Consequently, the influence of the nitrate concentration in the recirculating liquid on the H_2S *RE* was dependent of two factors: the H_2S *IL* and the *TLV*. A higher *TLV* level supplies a higher nitrate availability in the biofilm. *TLV* has usually been kept constant in anoxic BTFs between 10 [22] and 15 [23] $m \cdot h^{-1}$, but for a high H_2S *IL* it would be interesting to study the effect of this parameter. As in aerobic BTFs, where *TLV* is a key operational variable, in aerobic BTFs the regulation of *TLV* improves the oxygen mass transfer along the packed bed [24]. Fernández et al. [6] studied the effect of the *TLV* (2–20.5 $m \cdot h^{-1}$) on H_2S *RE* in an anoxic BTF, for H_2S *IL*s from 93 to 201 $gS \cdot m^{-3} \cdot h^{-1}$, packed with open pore polyurethane foam (the same support material as used in this study). It was found that there was no discernable influence for *TLV* values higher than 5 $m \cdot h^{-1}$ for H_2S *IL* values below 157 $gS \cdot m^{-3} \cdot h^{-1}$. However, at an H_2S *IL* of 201 $gS \cdot m^{-3} \cdot h^{-1}$ it was observed that *TLV* values below 15 $m \cdot h^{-1}$ produced a significant decrease in the H_2S *RE* from 92 to 85% at 4.5 $m \cdot h^{-1}$. On using polypropylene Pall rings [25] the optimal *TLV* was also 15 $m \cdot h^{-1}$ at high H_2S *IL* (>201 $gS \cdot m^{-3} \cdot h^{-1}$) although *TLV* did not have any effect at low H_2S *IL* (< 78.4 $gS \cdot m^{-3} \cdot h^{-1}$). Zeng et al. [3] studied the effect of *TLV* between 2.63 an 9.47 $m \cdot h^{-1}$ (H_2S *IL* < 86.92 $gS \cdot m^{-3} \cdot h^{-1}$) and achieved an efficient removal of H_2S for the lowest *TLV*, probably due to the larger height-diameter (*H/D*) ratio (10.9) and the higher *EBRT* (342 s). A high *H/D* ratio and a low *EBRT* improve the gas-liquid mass transfer [26] but increase the installation cost due to the higher pressure drop [27] and the higher volume of the packed bed.

The statistical results for the '*C model*' show the significance and high predictability of the regression model. The R-squared was 95.77%, the residual standard deviation was 0.1784 and the mean absolute error was 0.1224. The Durbin–Watson statistic was 1.00088 (p-value = 0.0001) and this shows a possible autocorrelation in the sample with a significance level of 5.0%. Moreover, the plot of residual versus predicted values (Figure 4a) does not show any patterns and we can assume a good correlation between the model prediction and the experimental results. The second-order polynomial model fitted with calibration data is represented by Equation (11).

$$C_{G,H2S} = -0.27436 + 0.0203426 \cdot TLV + 0.0203127 \cdot IL - 0.00935431 \cdot \left[N\text{--}NO_3^-\right]$$
$$-0.00120525 \cdot TLV^2 - 0.000824025 \cdot TLV \cdot IL + 0.000411032$$
$$\cdot TLV \cdot \left[N\text{--}NO_3^-\right] + 0.00000476214 \cdot IL^2 - 0.0000513469 \cdot IL$$
$$\cdot \left[N\text{--}NO_3^-\right] + 0.0000617476 \cdot \left[N\text{--}NO_3^-\right]^2 \tag{11}$$

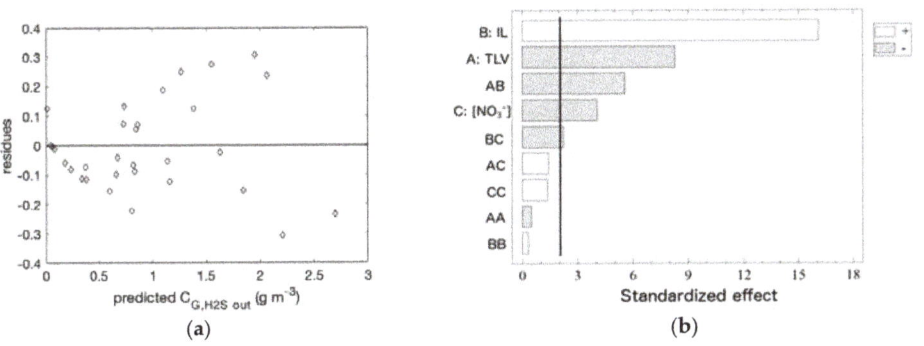

Figure 4. Analysis for the '*C model*': Residual versus predicted (**a**) and Pareto chart (**b**).

As can be seen in Figure 4b, the most influential factor on the outlet H_2S concentration was the H_2S *IL* (p-value $2.26 \cdot 10^{-8}$). Moreover, the *TLV* and nitrate concentration had a negative effect on the outlet H_2S concentration with p-values of $5.26 \cdot 10^{-14}$ and $5.34 \cdot 10^{-4}$, respectively. It is interesting to note that the interaction AB (*IL* and *TLV*) was significant (p-value $1.18 \cdot 10^{-5}$) and therefore the effect of one variable depended on the value of another. This behavior can be seen in Figure 3, where the *TLV* had an effect on the *RE* at high H_2S *IL* but not at low ones.

The response surface for the '*C model*' is shown in Figure 5. The model can be used to predict the factor limits to achieve a desired H_2S outlet concentration. Depending on the combustion engine company, the inlet H_2S limit is in the range 100–500 mg·Nm^{-3} (72–359 ppm$_V$ at 25 °C) [28]. Therefore, for a nitrate concentration of 35.5 mgN-NO$_3^-$·L^{-1} and *TLV* of 15.27 m h^{-1} the maximum H_2S *IL* would be 66.72 and 119.75 gS·m^{-3}·h^{-1} for outlet H_2S concentrations of 72 and 359 ppm$_V$, respectively.

The statistical results for the '*EC model*' show a higher significance and predictability of the regression model when compared with the '*C model*'. The R-squared was 99.63%, the residual standard deviation was 2.545 and the mean absolute error was 1.646. The Durbin–Watson statistic was 1.7041 (p-value = 0.0626), although the p-value was higher than 5% there were no traces of autocorrelation—as verified by checking the residual plot (Figure 6a). The second-order polynomial model fitted with calibration data is represented by Equation (12).

$$EC = 11.0413 - 2.30638 \cdot TLV + 0.844035 \cdot IL + 0.172511 \cdot \left[N\text{--}NO_3^-\right]$$
$$+0.090618 \cdot TLV^2 + 0.0225828 \cdot TLV \cdot IL - 0.00947479 \cdot TLV$$
$$\cdot \left[N\text{--}NO_3^-\right] - 0.00231493 \cdot IL^2 + 0.00184907 \cdot L \cdot \left[N\text{--}NO_3^-\right]$$
$$-0.00194474 \cdot \left[N\text{--}NO_3^-\right]^2 \tag{12}$$

As can be seen in Figure 6b, the most influential factor on the *EC* was the H_2S *IL* (p-value $1.81 \cdot 10^{-28}$). Moreover, the *TLV* and nitrate concentration had a positive effect on the *EC* with p-values of $2.37 \cdot 10^{-11}$ and $1.38 \cdot 10^{-5}$, respectively. In this case, the interactions AB (*IL* and *TLV*) and BC (*IL* and nitrate concentration) and the quadratic term (BB or IL^2) were more significant than the nitrate concentration. Therefore, H_2S *IL* and *TLV* had a greater effect on the *EC* than the nitrate concentration.

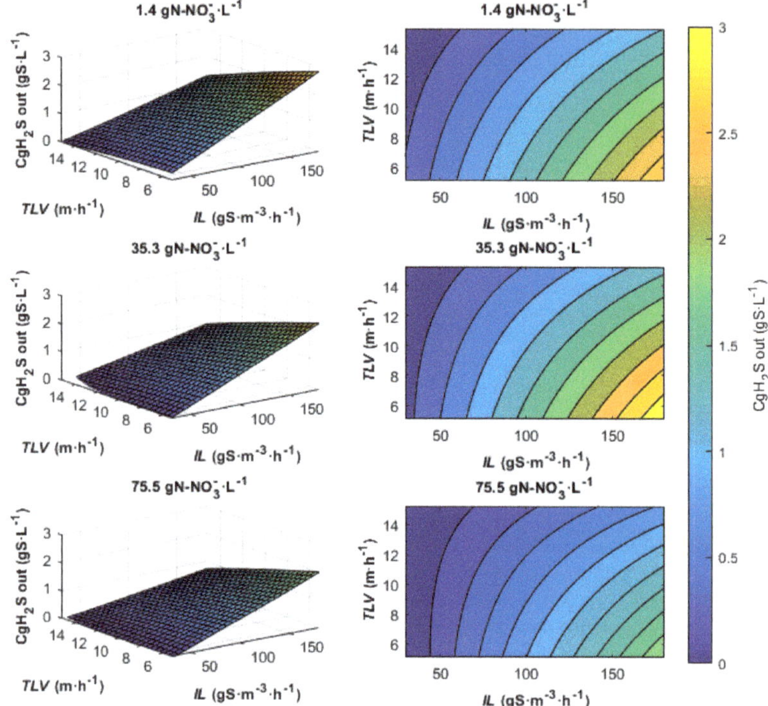

Figure 5. Response surfaces for the '*C model*'.

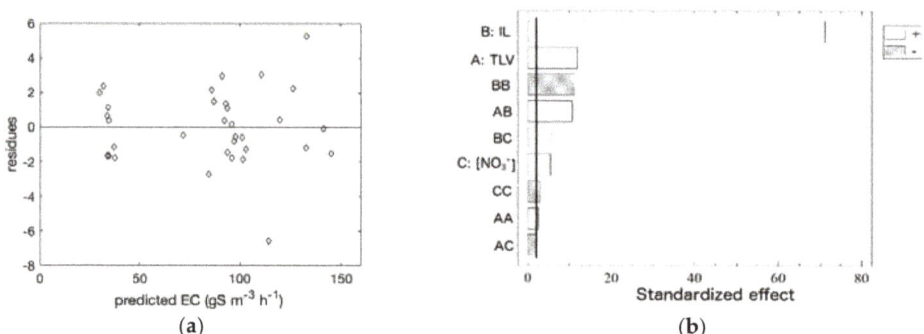

Figure 6. Analysis for the 'elimination capacity (EC) model': Residual versus predicted (**a**) and Pareto Chart (**b**).

The response surface for the '*EC model*' is shown in Figure 7. As expected, for high H_2S *IL* (>109.1 ± 11.7 gS·m^{-3}·h^{-1}) or high biogas flow rate (F_G > 3 m^3·h^{-1}) an increase in *EC* was observed when the nitrate concentration and *TLV* were increased.

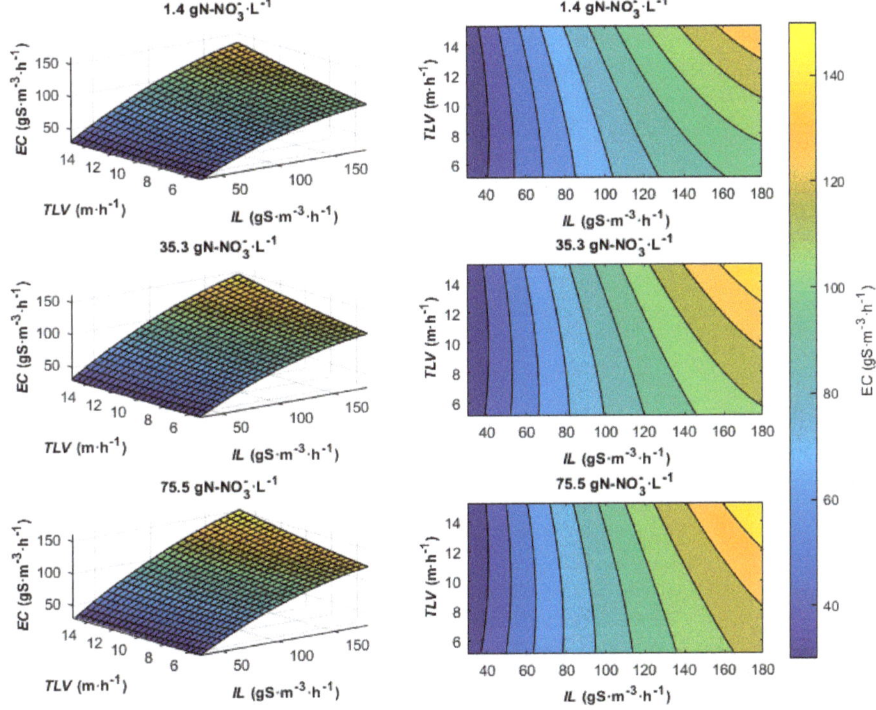

Figure 7. Response surface for the *'EC model'*.

A sensitivity analysis was performed by calculating the partial derivative in both models [29]. The maximum and minimum values for the variation of estimated variables corresponding to each factor are provided in Table 3. For the *'C model'* the maximum negative effect corresponded to a *TLV* of –0.1579 $(gS \cdot m^{-3})/(m \cdot h^{-1})$ and the maximum positive effect was for an H_2S *IL* of 0.0177 $(gS \cdot m^{-3})/(gS \cdot m^{-3} \cdot h^{-1})$. However for the *'EC model'* the maximum effects were due to *TLV* values of –1.252 $(gS \cdot m^{-3})/(m \cdot h^{-1})$ and 4.3303 $(gS \cdot m^{-3})/(m \cdot h^{-1})$.

Table 3. Sensitivity analysis.

Factor	*'C Model'*		*'EC Model'*	
	Maximum	Minimum	Maximum	Minimum
TLV	0.0078	–0.1579	4.3303	–1.252
IL	0.0177	0.0045	1.1553	0.1657
[N-NO$_3$$^-$]	0.0038	–0.159	0.4363	–0.1801

3.2. Ottengraf's Model

The concentration profiles along the bed height were analyzed using Ottengraf's model, without nitrate concentration limitations, for H_2S *IL* values between 33 and 176 $gS \cdot m^{-3} \cdot h^{-1}$ and *TLV* values between 5.09 and 15.27 $m \cdot h^{-1}$. The linear adjustments are provided in Table 4 according to the following simplifications: controlled by zero-order diffusion, zero-order kinetic and first-order kinetic. The behavior of the concentration profile for 5.09 $m \cdot h^{-1}$ and 130 and 170 $gS \cdot m^{-3} \cdot h^{-1}$ could be explained using a zero-order simplification for diffusion or kinetic. However, for all other conditions the first-order kinetic simplification can be applied. The kinetic constant was in the range between 0.0025 and 0.0092 s^{-1}.

Table 4. Adjustments using Ottengraf's model.

IL (gS·m^{-3}·h^{-1})	TLV (m·h^{-1})	Zero-Order Diffusion		Zero-Order Kinetic		First-Order Kinetic	
		r^2	k (s^{-1})	r^2	k (s^{-1})	r^2	k (s^{-1})
36	5.09	0.62	0.0018	0.3	0.012	0.95	0.008
35	10.18	0.57	0.0018	0.26	0.012	0.92	0.0087
33	15.27	0.52	0.0018	0.22	0.011	0.93	0.0092
130	5.09	0.95	0.0011	0.94	0.011	0.93	0.0031
108	10.18	0.91	0.0012	0.82	0.01	0.93	0.0036
104	15.27	0.93	0.0014	0.77	0.01	0.96	0.0057
170	5.09	0.97	0.0009	0.96	0.0079	0.96	0.0025
176	10.18	0.88	0.001	0.79	0.0091	0.93	0.0029
167	15.27	0.86	0.0011	0.74	0.0092	0.95	0.0034

To our knowledge, Ottengraf's model has not be applied to biogas desulfurization, although studies on H$_2$S removal from air have been modeled using Ottengraf's model [30–33].

Jin et al. [32] found that the zero-order kinetic limitation described the outlet H$_2$S concentration (TLV of 0.62 m·h^{-1} and a maximum H$_2$S IL around 30 gS·m^{-3}·h^{-1}). Oyarzún et al. [30] applied zero-order diffusion equations for an inlet H$_2$S concentration below Ks (Monod saturation constant) and zero-order kinetic equation and inlet concentration above Ks. The microbial kinetic of the biotrickling filter presented in this work can be described by a Haldane model [15], with an affinity constant for sulfide (Ks) of 8.4 gS·m^{-3} and a gas concentration in equilibrium of 1.28 gS·m^{-3}. This concentration is considerably lower than the minimum inlet concentration employed in this work (5.55 gS·m^{-3}). Therefore, the study was carried out at an inlet H$_2$S concentration higher than Ks with a first-order kinetic as obtained by Oyarzún et al. [30].

4. Conclusions

The influence of the nitrate concentration was dependent on TLV and H$_2$S IL, with its influence increasing for lower TLV and higher H$_2$S IL. The empirical models obtained by the response surface methodology for the factorial design of three factors at three levels (3^3) were able to predict the outlet H$_2$S concentration and the EC with a R-squared of 95.77% and 99.63% without autocorrelation. The most influential factors on the outlet H$_2$S concentration and EC were the H$_2$S IL and TLV, with the nitrate concentration being less significant. For biogas use in a CHP system the maximum H$_2$S IL should be between 66.72 and 119.75 gS·m^{-3}·h^{-1} (TLV of 15.27 m·h^{-1} and nitrate concentration of 35.5 mgN-NO$_3^-$·L^{-1}). Moreover, Ottengraf's model was applied successfully considering a first-order kinetic limitation simplification with an R-squared above 0.92.

Author Contributions: Research design, D.C. and M.R.; methodology, F.A. and M.R.; investigation, F.A.; writing—original draft preparation, F.A.; writing—review and editing, M.R.; supervision, D.C. and M.R.; funding acquisition, D.C. and M.R.

Funding: This research was funded by MINISTERIO DE ECONOMÍA Y COMPETITIVIDAD, grant number CTM2009-14338-C03-02 and the Research Result Transfer Office of the University of Cádiz, grant number PROTO-05-2010.

Acknowledgments: The authors wish to express sincere gratitude to the WWTP "UTE EDAR Bahía de Cádiz" for allowing us to install the biotrickling filter in their plant.

Conflicts of Interest: The authors declare no conflict of interest.

References

1. European Commission. *A Clean Planet for All a European Strategic Long-Term Vision for a Prosperous, Modern, Competitive and Climate Neutral Economy*; COM (2018) 773 Final; European Commission: Brussels, Belgium, 2018; pp. 1–25.
2. Cano, P.I.; Colón, J.; Ramírez, M.; Lafuente, J.; Gabriel, D.; Cantero, D. Life cycle assessment of different physical-chemical and biological technologies for biogas desulfurization in sewage treatment plants. *J. Clean. Prod.* **2018**, *181*, 663–674. [CrossRef]
3. Zeng, Y.; Luo, Y.; Huan, C.; Shuai, Y.; Liu, Y.; Xu, L.; Ji, G.; Yan, Z. Anoxic biodesulfurization using biogas digestion slurry in biotrickling filters. *J. Clean. Prod.* **2019**, *224*, 88–99. [CrossRef]
4. Valle, A.; Fernández, M.; Ramírez, M.; Rovira, R.; Gabriel, D.; Cantero, D. A comparative study of eubacterial communities by PCR-DGGE fingerprints in anoxic and aerobic biotrickling filters used for biogas desulfurization. *Bioprocess Biosyst. Eng.* **2018**, *41*, 1165–1175. [CrossRef] [PubMed]
5. Montebello, A.M.; Fernández, M.; Almenglo, F.; Ramírez, M.; Cantero, D.; Baeza, M.; Gabriel, D. Simultaneous methylmercaptan and hydrogen sulfide removal in the desulfurization of biogas in aerobic and anoxic biotrickling filters. *Chem. Eng. J.* **2012**, *200–202*, 237–246. [CrossRef]
6. Fernández, M.; Ramírez, M.; Gómez, J.; Cantero, D. Biogas biodesulfurization in an anoxic biotrickling filter packed with open-pore polyurethane foam. *J. Hazard. Mater.* **2014**, *264*, 529–535. [CrossRef] [PubMed]
7. Fortuny, M.; Baeza, J.; Gamisans, X.; Casas, C.; Lafuente, J.; Deshusses, M.; Gabriel, D. Biological sweetening of energy gases mimics in biotrickling filters. *Chemosphere* **2008**, *71*, 10–17. [CrossRef]
8. Wainwright, J.; Mulligan, M. *Environmental Modelling: Finding Simplicity in Complexity*, 2nd ed.; John Wiley & Sons, Inc.: Hoboken, NJ, USA, 2013; ISBN 978-0-470-74911-1.
9. Ottengraf, S.; Rehm, H.; Reed, G. Exhaust gas purification. In *Biotechnology—A Comprehensive Treatise*; Verlag Chemie: Weinheum, Germany, 1986; Volume 8, pp. 426–452.
10. Shareefdeen, Z.; Baltzis, B.; Oh, Y.; Bartha, R. Biofiltration of Methanol Vapor. *Biotechnol. Bioeng.* **1993**, *41*, 512–524. [CrossRef]
11. Baltzis, B.; Wojdyla, S.; Zarook, S. Modeling biofiltration of VOC mixtures under steady-state conditions. *J. Environ. Eng.* **1997**, *123*, 599–605. [CrossRef]
12. Soreanu, G. Insights into siloxane removal from biogas in biotrickling filters via process mapping-based analysis. *Chemosphere* **2016**, *146*, 539–546. [CrossRef]
13. Soreanu, G.; Falletta, P.; Béland, M.; Edmonson, K.; Ventresca, B.; Seto, P. Empirical modelling and dual-performance optimisation of a hydrogen sulphide removal process for biogas treatment. *Bioresour. Technol.* **2010**, *101*, 9387–9390. [CrossRef]
14. López, L.; Dorado, A.; Mora, M.; Gamisans, X.; Lafuente, J.; Gabriel, D. Modeling an aerobic biotrickling filter for biogas desulfurization through a multi-step oxidation mechanism. *Chem. Eng. J.* **2016**, *294*, 447–457. [CrossRef]
15. Almenglo, F.; Ramírez, M.; Gómez, J.; Cantero, D.; Gamisans, X.; Dorado, A. Modeling and control strategies for anoxic biotrickling filtration in biogas purification. *J. Chem. Technol. Biotechnol.* **2016**, *91*, 1782–1793. [CrossRef]
16. Almenglo, F.; Ramírez, M.; Gómez, J.; Cantero, D. Operational conditions for start-up and nitrate-feeding in an anoxic biotrickling filtration process at pilot scale. *Chem. Eng. J.* **2016**, *285*, 83–91. [CrossRef]
17. Clesceri, L.S.; Greenberg, A.; Eaton, A. *Standard Methods for the Examination of Water and Waste Water*, 20th ed.; American Public Health Association, American Water Works Association, Water Environment Federation: Washington, DC, USA, 1999; ISBN 0875532357.
18. Veljković, V.B.; Veličković, A.V.; Avramović, J.M.; Stamenković, O.S. Modeling of biodiesel production: Performance comparison of box–Behnken, face central composite or full factorial design. *Chin. J. Chem. Eng.* **2018**. [CrossRef]
19. Montgomery, D.C. *Design and Analysis of Experiments*, 5th ed.; John Wiley & Sons, Inc.: Hoboken, NJ, USA, 2001; p. 208. ISBN 0-471-31649-0.
20. Ottengraf, S.; Oever, V.A. Kinetics of organic compound removal from waste gases with a biological filter. *Biotechnol. Bioeng.* **1983**, *25*, 3089–3102. [CrossRef] [PubMed]
21. Onda, K.; Takeuchi, H.; Okumoto, Y. Mass transfer coefficients between gas and liquid phases in packed columns. *J. Chem. Eng. Jpn.* **1968**, *1*, 56–62. [CrossRef]

22. López, L.R.; Brito, J.; Mora, M.; Almenglo, F.; Baeza, J.A.; Ramírez, M.; Lafuente, J.; Cantero, D.; Gabriel, D. Feedforward control application in aerobic and anoxic biotrickling filters for H_2S removal from biogas. *J. Chem. Technol. Biotechnol.* **2018**, *93*, 2307–2315. [CrossRef]

23. Brito, J.; Valle, A.; Almenglo, F.; Ramírez, M.; Cantero, D. Progressive change from nitrate to nitrite as the electron acceptor for the oxidation of H_2S under feedback control in an anoxic biotrickling filter. *Biochem. Eng. J.* **2018**, *139*, 154–161. [CrossRef]

24. López, L.R.; Bezerra, T.; Mora, M.; Lafuente, J.; Gabriel, D. Influence of trickling liquid velocity and flow pattern in the improvement of oxygen transport in aerobic biotrickling filters for biogas desulfurization. *J. Chem. Technol. Biotechnol.* **2016**, *91*, 1031–1039. [CrossRef]

25. Fernández, M.; Ramírez, M.; Pérez, R.; Gómez, J.; Cantero, D. Hydrogen sulphide removal from biogas by an anoxic biotrickling filter packed with Pall rings. *Chem. Eng. J.* **2013**, *225*, 456–463. [CrossRef]

26. Ordaz, A.; Figueroa-González, I.; San-Valero, P.; Gabaldón, C.; Quijano, G. Effect of the height-to-diameter ratio on the mass transfer and mixing performance of a biotrickling filter. *J. Chem. Technol. Biotechnol.* **2018**, *93*, 121–126. [CrossRef]

27. Lebrero, R.; Gondim, A.; Pérez, R.; García-Encina, P.A.; Muñoz, R. Comparative assessment of a biofilter, a biotrickling filter and a hollow fiber membrane bioreactor for odor treatment in wastewater treatment plants. *Water Res.* **2014**, *49*, 339–350. [CrossRef] [PubMed]

28. Ramírez, M.; Gómez, J.; Cantero, D.; Ramírez, M.; Gómez, J.; Cantero, D. Biogas: Sources, Purification and Uses. In *Hydrogen and Other Technologies*; Studium Press LLC: Houston, TX, USA, 2015; Volume 11, pp. 296–323. ISBN 978-1-626990-72-2.

29. Koda, M.; Dogru, A.H.; Seinfeld, J.H. Sensitivity analysis of partial differential equations with application to reaction and diffusion processes. *J. Comput. Phys.* **1979**, *30*, 259–282. [CrossRef]

30. Oyarzún, P.; Arancibia, F.; Canales, C.; Aroca, G.E. Biofiltration of high concentration of hydrogen sulphide using *Thiobacillus thioparus*. *Process Biochem.* **2003**, *39*, 165–170. [CrossRef]

31. Jaber, M.; Couvert, A.; Amrane, A.; Rouxel, F.; Cloirec, P.; Dumont, E. Biofiltration of H_2S in air—Experimental comparisons of original packing materials and modeling. *Biochem. Eng. J.* **2016**, *112*, 153–160. [CrossRef]

32. Jin, Y.; Veiga, M.C.; Kennes, C. Effects of pH, CO_2, and flow pattern on the autotrophic degradation of hydrogen sulfide in a biotrickling filter. *Biotechnol. Bioeng.* **2005**, *92*, 462–471. [CrossRef] [PubMed]

33. Shareefdeen, Z.M.; Ahmed, W.; Aidan, A. Kinetics and Modeling of H_2S Removal in a Novel Biofilter. *Adv. Chem. Eng. Sci.* **2011**, *1*, 72–76. [CrossRef]

 *chemengineering*

Article

Evaluation of Biogas Biodesulfurization Using Different Packing Materials

Samir Prioto Tayar [1], Renata de Bello Solcia Guerrero [2], Leticia Ferraresi Hidalgo [1] and **Denise Bevilaqua [1],***

1   Department of Biochemistry and Chemistry Technology, Institute of Chemistry, São Paulo State University (UNESP), Araraquara 14800-900, Brazil; samir.ptayar@gmail.com (S.P.T.); leticia.ferraresi@gmail.com (L.F.H.)
2   Biological Process Laboratory, Center for Research, Development and Innovation in Environmental Engineering, São Carlos School of Engineering, University of São Paulo (EESC/USP), Av. João Dagnone, 1100, São Carlos 13.563-120, Brazil; rsolcia@gmail.com
*   Correspondence: denise.bevilaqua@unesp.br; Tel.: +55-16-3301-9677

Received: 18 December 2018; Accepted: 28 February 2019; Published: 8 March 2019

Abstract: The packing material selection for a bioreactor is an important factor to consider, since the characteristics of this material can directly affect the performance of the bioprocess, as well as the investment costs. Different types of low cost packing materials were studied in columns to reduce the initial and operational costs of biogas biodesulfurization. The most prominent (PVC pieces from construction pipes) was applied in a bench-scale biotrickling filter to remove the H_2S of the biogas from a real sewage treatment plant in Brazil, responsible for 90 thousand inhabitants. At the optimal experimental condition, the reactor presented a Removal Efficiency (RE) of up to 95.72% and Elimination Capacity (EC) of 98 $gS·m^{-3}·h^{-1}$, similar to open pore polyurethane foam, the traditional material widely used for H_2S removal. These results demonstrated the high potential of application of this packing material in a full scale considering the robustness of the system filled with this support, even when submitted to high sulfide concentration, fluctuations in H_2S content in biogas, and temperature variations.

Keywords: packing material; PVC; open-pore polyurethane foam; PET; Teflon; biotrickling filter; hydrogen sulfide elimination; H_2S

---

1. Introduction

Biogas, generated in wastewater treatment plants, typically composed of methane (CH_4), carbon dioxide (CO_2), and traces of hydrogen sulfide (H_2S), is often burned in flares to minimize its contribution to the greenhouse effect [1]. However, in the current global scenario, its commercial value has been increasing due to its high-energy content. The use of this versatile and renewable energy source can bring high cost savings [2]. A limiting factor for the use of this biogas is the wide variety of contaminants present in its composition, such as sulfur compounds, siloxanes, hydrocarbons, and halogenated organic compounds, of which H_2S is the most harmful for energy conversion equipment, due to its corrosive character.

Bed clogging is one of the biggest problems of biotrickling filters, being caused by solid accumulation (biomass and elemental sulfur), and limiting high treatment rates [3]. Problems with elemental sulfur accumulation have been observed since one of the first works described for H_2S removal from biogas in a biotrickling filter (under aerobic conditions) [4]. Due to this, a wide variety of packing materials has been studied in order to overcome this problem and improve the biofilter performance; among them are open pore polyurethane foam (OPUF), polyester fibers, pall rings, porous lava rock, activated carbon, glass beads, and perlite. On the other hand, the choice of the packing material is also related to the economic viability of the biofilters, since materials with a low purchase cost and low pressure drop (low resistance to gas flow) can significantly decrease the operating cost. In this point of view, the use of low-cost packing materials such as expanded schist (inorganic) and cellular concrete waste was also studied for biogas biodesulfurization [5].

In biotrickling filters, the gas flow passes through an inert packing material to which the microbial community attaches. Considering that different types of forces (electrostatic and hydrophobic interactions, and covalent and partial covalent bond formation, among others), are involved in the microbial attachment to a packing material [6], the material surface properties can implicate in different biofilm formation, resulting in different performance of bioreactors. Therefore, the characteristics of the packing material play an important role in the sulfide elimination from biogas (Table 1) and some factors should be observed to choose the best support material:

(a)   High surface area for biofilm growth and mass transfer;
(b)   Hydrophobicity;
(c)   Mechanical, chemical, and biological resistance;
(d)   Low pressure drop, especially considering pilot operation

Table 1. Packing material characteristics.

Packing Material	Specific Surface Area ($m^2 \cdot m^{-3}$)	Density ($kg \cdot m^{-3}$)	Porosity (%)	Reference
Lava rocks	200 ± 50	-	-	[7]
Plastic fibers	650 ± 50	-	-	[7]
Open pore polyurethane foam	600	35	97	[8]
Polypropylene pall rings	320	110	88	[9]
Metallic Pall rings	515	520	-	[10]
Honeycomb	620	-	88	[11]

Open pore polyurethane foam is a commercial packing material developed especially for biotrickling filtration [12]. Although conventional polyurethane foam is hydrophobic, its properties are extensive and can vary due to the starting molecules and reaction conditions of manufacture [13]. According to Lisiecki et al. [14], open pore polyurethane foam (TM25450) is formed by thermal compression of conventional foam which leads the cell walls to collapse. The advantages of OPUF are its high porosity, suitable pore size, low density, high specific surface area and reasonable resistance to compaction [8].

The first works carried out using OPUF were developed to treat odorous air. Gabriel et al. [8] proposed the use of OPUF foam as packing support based on the successful results obtained by Loy et al. [15].

Under anoxic conditions, Fernández et al. [16] developed one of the first works using OPUF cubes ($8\ cm^3$), obtaining a critical EC of 60 gS-H_2S$\cdot m^{-3} \cdot h^{-1}$ (empty bed residence time (EBRT) of 240 s). Using the same system (packed bed volume of 2.375 L), Montebello et al. [17] showed the simultaneous removal of H_2S and methylmercaptan (CH_3SH) from biogas to be feasible, however, loads higher than 100 gS-H_2S$\cdot m^{-3} \cdot h^{-1}$ negatively affected the CH_3SH removal due to competition. The maximum elimination capacity achieved was 140 gS-H_2S$\cdot m^{-3} \cdot h^{-1}$.

Fernández et al. [18] operated a laboratory scale biotrickling filter for 620 days and demonstrated that the optimal conditions were: Sulfate concentration below 33 g$\cdot L^{-1}$, pH between 7.3–7.5, temperature of 30 °C, and trickling liquid velocity (TLV) higher than 4.6 m$\cdot h^{-1}$. Higher critical elimination capacity was observed under a nitrate programmed feeding regime (130 gS$\cdot m^{-3} \cdot h^{-1}$, RE 99%, EBRT 2.4 min) when compared to a manual feeding regime (99.8 gS$\cdot m^{-3} \cdot h^{-1}$, RE < 99%, EBRT 3.4 min). The maximum elimination capacity was 170 gS$\cdot m^{-3} \cdot h^{-1}$ for both regimes. Comparatively, Guerrero and Bevilaqua [19] observed optimum temperatures between 31–42 °C and EBRTs from 2.9 to 6.2 min for H_2S removal from the biogas generated from an up flow anaerobic sludge blanket reactor (UASB) at the wastewater treatment plant of a brewery.

On the other hand, for a pilot-scale biotrickling filter, Almenglo et al. [20] observed the best performance under counter-current flow, achieving a maximum elimination capacity of 140 gS·m^{-3}·h^{-1}. Almenglo et al. [21] recommended some important guidelines to startup the biotrickling filter in order to avoid sulfide accumulation in the early stages of reactor operation: An inlet load (IL) around 100 gS·m^{-3}·h^{-1} and pH of 6.8 to decrease the solubility of the sulfide.

Pall rings are widely used in chemical scrubbers due to their high free volume, low-pressure drop values, and uniform gas-liquid contact. Due to these characteristics, it is possible to transform conventional scrubbers into biotrickling filters just carrying out microorganism's immobilization. Polypropylene Pall rings have low specific surface area when compared with open pore polyurethane foam and other packing materials (Table 1), however, according to Fernández et al. (2013), this characteristic can minimize the pressure loss due to biomass and sulfur accumulation. Pall rings are a hydrophobic packing material [22].

The conversion from scrubber to biotrickling filter was first applied to air pollution control; however, in recent decades, the use of Pall rings as packing material has extended to biotrickling filters for H$_2$S removal from biogas. Under aerobic conditions, Tomas et al. [23] achieved a maximum elimination capacity of 170 gS-H$_2$S·m^{-3}·h^{-1} in a full-scale biotrickling filter (gas contact time of 180 s, pH 2.6–2.7). The biofilter was composed of four modular sections with an inner diameter of 1.4 m and height of 8 m. Montebello et al. [10] operated a biotrickling filter packed with metallic Pall rings for approximately two years, treating synthetic biogas with a 2000 ppmv H$_2$S concentration. Under neutral pH, the maximum and critical EC were approximately 100 gS·m^{-3}·h^{-1} and under oxygen appropriate load it was possible to minimize the elemental sulfur formation. This behavior was not observed under acidic pH. In addition, under acidic pH, deterioration of the packing material was observed. Under anoxic conditions, Fernández et al. [9] achieved 99% H$_2$S removal efficiency under inlet loads lower than 120 gS·m^{-3}·h^{-1} in an anoxic biotrickling filter (working volume of 2.4 L), using controlled nitrate feeding by oxidize-reduction potential (ORP).

López et al [24] studied the main parameters involved in the oxygen mass transfer efficiency in an aerobic biotrickling filter, in order to reduce the elemental sulfur production under high H$_2$S loads. The trickling liquid velocity and co-current flow showed to be better to manipulate when compared to air supply flow rate and counter current flow mode, increasing 10% the EC and 9% the selectivity to sulfate as product of the H$_2$S oxidation under 283.8 gS-H$_2$S·m^{-3}·h^{-1}. López et al [25] also used the feedforward control through the trickling liquid velocity and observed a reduction of 68.4% of the maximum outlet H$_2$S concentration and the sulfate selectivity improved 100.6 $\pm$ 5.0%. A biotrickling filter packed with Pall rings was used by López et al. [26] to develop, calibrate, and validate a dynamic model to describe the main processes involved in the H$_2$S removal from biogas (high loads). The model was capable to predict the biotrickling operation, besides being able to describe the main products of the H$_2$S oxidation.

Other packing materials have been used under aerobic conditions. Montebello et al. [27] evaluated the performance of a biotrickling filter packed with HD-QPAC (volume of 2.15 L) under IL from 51 to 215 gH$_2$S·m^{-3}·h^{-1} and observed a maximum elimination capacity of 201 gH$_2$S·m^{-3}·h^{-1} and maximum RE of 100% (EBRT of 180 s). The decrease in the O$_2$/H$_2$S ratio resulted in an increase in S^0 production. Fortuny et al. [28] also used HD-QPAC as packing material in a biotrickling filter (total volume of 2 L). An important effect of the EBRT on the RE was observed when the EBRT decreased from 120 s (97.7 $\pm$ 0.3%) to 30 s (39.7 $\pm$ 0.9%).

Qiu and Deshusses [11] showed a promising alternative to the use of conventional packing material. The authors evaluated the use of 3D-printed honeycomb monolith, composed of 19 hexagonal channels, in order to reduce bed-clogging problems under high H$_2$S concentration through the presence of connected and straight channels. The elimination capacity exceeded 120 gS·m^{-3}·h^{-1} at an H$_2$S/O$_2$ ratio of 1:2. Elemental sulfur was obtained as the predominant end-product, with accumulation in the bed. However, the bed pigging was shown to be efficient for removing elemental sulfur and excess biomass.

Vikromvarasiri et al. [29] studied a biotrickling filter filled with random packing media (working volume of 1 L) and inoculated with *Halothiobacillus neapolitanus* NTV01 (HTN), isolated from activated sludge. Air was supplied as the final electron acceptor. Different operational parameters were compared under short-term and long-term operation. The relationship between IL and EC was higher in the long-term (0.931) when compared to short-time operation (0.915). The maximum elimination capacity obtained was 78.57 $gH_2S \cdot m^{-3} \cdot h^{-1}$ for an IL of 85.25 $gH_2S \cdot m^{-3} \cdot h^{-1}$.

Recently, Jaber et al. [5] proposed the use of a biofilter packed with expanded schist (inorganic) and cellular concrete waste (recycled mineral waste), low cost materials, to treat H_2S from biogas. Both materials demonstrated low pressure drops, which is very desirable. The maximum elimination capacity obtained for the expanded schist was 30.3 $g \cdot m^{-3} \cdot h^{-1}$ and 25.2 $g \cdot m^{-3} \cdot h^{-1}$ for the cellular concrete waste.

In recent decades, some studies have been carried out to demonstrate the viability of retrofitting existing chemical scrubbers to full-scale biotrickling filters for H_2S control. Gabriel et al. [8] and Gabriel and Deshusses [30] demonstrated that biotrickling filters can replace chemical scrubbers successfully, being safe and economical. Nevertheless, Gabriel et al. [8] estimated that the expenditure on packing material is approximately $500–1000/m^3, representing a large percentage of the costs. Tomas et al. [23] compared the investment costs involved in H_2S removal using chemical and biological treatment and showed an investment of $52,000 for the biological treatment (including reactor, blower, pump, and packing material) compared to $8700 for the chemical oxidation treatment. This evidence reinforces the need to search for new packing materials aimed at reducing investment costs. Recently, Cano et al. [31] compared the life cycle of different technologies such as aerobic biotrickling filtration, anoxic biotrickling filtration, caustic chemical scrubbing and absorption on impregnated activated carbon, and it was included the analyses of the capital expenditures. The results showed that a biotrickling filter made with fiberglass reinforced plastic (total volume 10 m^3, packing bed volume 5.3 m^3, diameter 0.9 m) can cost €17,090, compared to €5224 of a scrubber made of the same material (total volume 1.7 m^3, packing bed volume 1.1 m^3, diameter 0.7 m).

In this work, the immobilization of biomass on different low-cost packing materials, such as, PET, PVC and Teflon, was studied and the material that showed the best thiosulfate removal efficiency (PVC) was evaluated in a biotrickling filter for H_2S removal from real biogas. The effects of the EBRT, temperature and IL were studied.

2. Materials and Methods

2.1. Inoculum and Packing Material

The anaerobic sludge used in this work was obtained from the Matão Sewage Treatment Plant (São Paulo, Brazil).

Strips of Polyvinyl Chloride (PVC) obtained from Tigre® building pipes (Ref. 10121744 from manufacturer, Tigre, Brazil), Polyethylene Terephtalate (PET) from common soda bottle, Polytetrafluoroethylene (Teflon®, Teflon, Brazil) dowels cut into strips and OPUF (Filtren TM25450, Recticel Iberica, Spain) were used in order to evaluate the potential of low-cost carrier materials (Figure 1) for the immobilization of sulfur-oxidizing microorganism. The surface area of each packing material was analyzed based on the Brunauer, Emmett and Teller theory by the Center for Materials Characterization and Development (Ufscar, São Carlos, Brazil) (Table 2).

Table 2. Packing material characteristics [32].

Material	Surface Area ($m^2 \cdot g^{-1}$)
PVC	0.432
PET	0.443
Teflon	0.909
OPUF	6.694

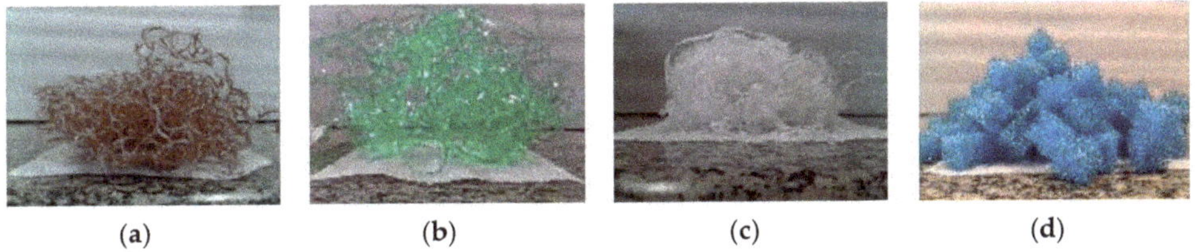

(a)　　　　　　(b)　　　　　　(c)　　　　　　(d)

Figure 1. Packing materials: (**a**) PVC; (**b**) PET; (**c**) Teflon®; (**d**) OPUF.

2.2. Experimental Set-Up of Laboratory-Columns Packed with Different Low Cost Materials

Four glass columns (active height of 280 mm, inner diameter of 60 mm, and working volume of 792 mL) packed with PVC, PET, Teflon and OPUF (Figure 2) were inoculated with 120 mL of anaerobic sludge and 120 mL of culture medium DSMZ 113, which has thiosulfate as a substrate, nitrate as the final electron acceptor, and bicarbonate as an inorganic carbon source. The thiosulfate consumption was monitored and after substrate depletion, 50% of the trickling solution was drawn off and fresh DSMZ 113 medium was added in the following 9 cycles. Different initial concentrations of thiosulfate (2.5, 5.0 and 10 g·$[S-S_2O_3{}^{2-}]\cdot L^{-1}$) were applied. The system was operated under batch mode for 130 days (total of 10 cycles) at pH 7.0 by adding NaOH, temperature 35 °C and trickling medium at a flow rate of 500 mL·h^{-1}. The immobilized biomass was determined as gram of protein per gram of dry support sampled in the end of the experiment.

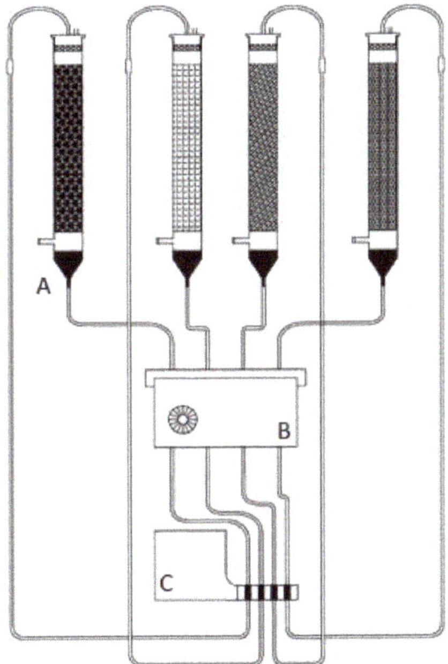

Figure 2. Experimental system: (**A**) Glass columns packed with different support materials; (**B**) water bath that maintained the temperature of the trickling medium at 35 °C; and (**C**) four-channel peristaltic pump.

2.3. Experimental Set-Up of a Laboratory-Scale Biotrickling Filter

The H$_2$S elimination from the biogas produced by a Sewage Treatment Plant (Matão, São Paulo, Brazil) was performed using a laboratory-scale biotrickling filter made of glass (active height of

456 mm, inner diameter of 93 mm and working volume bed of 3 L) filled with PVC pieces (Figure 3), obtained from construction pipes, as packing material. The experiments were carried out for 111 days with continuous biogas supply from the up flow anaerobic sludge blanket reactor (UASB).

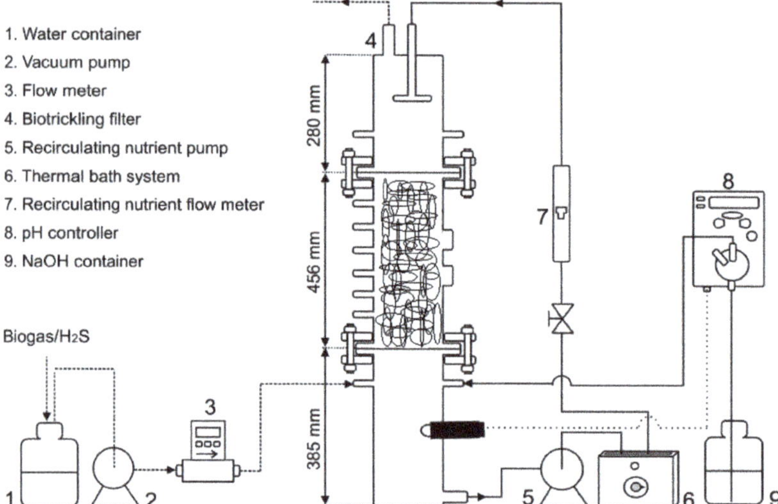

Figure 3. Representation of the biotrickling filter, adapted from Guerrero and Bevilaqua [19].

The first step of reactor operation consisted of inoculation and biofilm development providing H_2S containing biogas as the energy source (1246 ± 305 ppmv). Biofilm development was achieved after 43 days of operation and the degree of immobilization was determined by the stability of the substrate consumption [33]. The effects of the parameters, inlet load (from 8 to 108 $gS·m^{-3}·h^{-1}$), empty bed residency time (4.8, 2.4, 1.6 min), and temperature (from 24 to 40 °C), were studied in the following 68 days of operation.

2.4. Analytical Techniques

The H_2S concentration in the gas phase was measured using GasAlertMicro 5 Series—BW Technologies. The thiosulfate determination was carried out by an iodometric method [34]. Sulfate concentration was determined using a turbidimetric method and nitrate and nitrite concentrations were analyzed by an ultraviolet spectrophometric method and a colorimetric method, respectively [35]. The amount of biomass immobilized on the packing material was estimated via determining protein by Lowry method [35].

3. Results and Discussion

3.1. Evaluation of Low-Cost Packing Materials

All four packing material tested showed similar thiosulfate RE (between 82 and 86%) at the end of first cycle of rector operation. However, at the end of the last cycles, PVC (96.67%) and PET (96.43%) presented better results compared to Teflon (25.37%), which suffered bed compaction. Additionally, the RE obtained using PVC and PET almost reached the value obtained for OPUF (99.17%). On the other hand, the biomass protein quantification showed that the mass of protein per mass of support material was very similar for PVC (14.9 $mg·g^{-1}$) and Teflon (14.87 $mg·g^{-1}$) (Figure 4), however, as mentioned before, Teflon became compacted during the experiment. PET had a lower amount of protein (8.40 $mg·g^{-1}$) and all support materials presented lower biomass protein when compared with open-pore polyurethane foam (26.59 $mg·g^{-1}$).

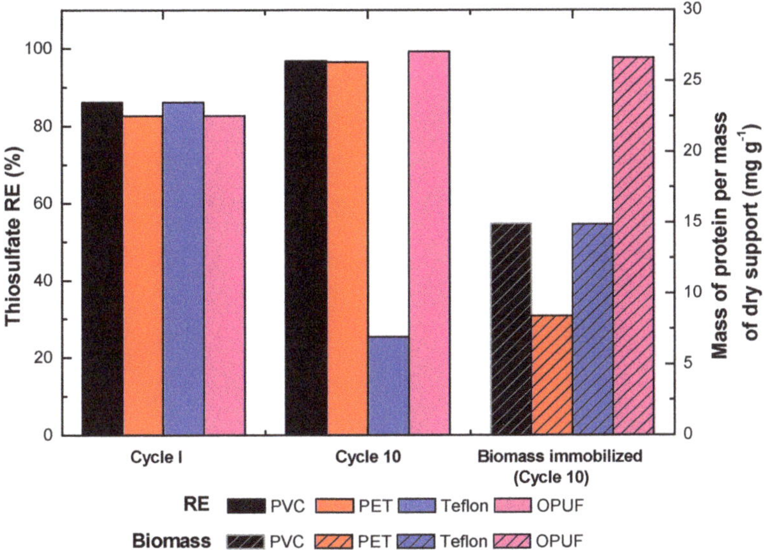

Figure 4. Thiosulfate Removal efficiency and biomass immobilized on the packing material after 10 cycles.

The cost to fill the working volume of the columns were: OPUF ($0.67 dollar), PVC ($0.21 dollar), PET ($0.0 dollar because it was from cycling residue) and Teflon ($0.36 dollar). Comparing the results, it was possible to suppose that PVC could provide better robustness to the system than PET, in case of adverse conditions, and other important factor is the cost of the PVC, 42% cheaper than Teflon and 69% cheaper than OPUF. For this reason, it was selected as packing material for the next experiment using a biotrickling filter to treat real biogas in bench scale.

3.2. Biotrickling Filter Operation

During the first 70 days of operation (including the inoculation and biofilm development step) the EBRT applied was 4.8 min and the inlet load provided to the reactor was 15.5 ± 3.09 gS·m^{-3}·h^{-1}. In these conditions a removal efficiency (RE) of $80.06 \pm 13.81\%$ was observed (Figure 5), lower than experiments with lower EBRTs of 2.4 min (RE = $71.89 \pm 14.33\%$; IL = 39.9 gS·m^{-3}·h^{-1}) and 1.6 min (RE = $89.83 \pm 14.97\%$; 57.9 gS·m^{-3}·h^{-1}). These results reflected the impact of the biofilm development step to elimination capacity in this operating condition, showing that, after stabilization, the biotrickling filter was capable of achieving a better RE even with a lower retention time. After stabilization, the system showed a high RE despite the low EBRT (1.6 min) and high H$_2$S concentration (1832 ± 295 ppmv), demonstrating that this bioreactor requires good biofilm development to achieve high EC in a real treatment which presents fluctuations over time.

The elimination capacity of the system obtained under different EBRTs was very high (Figure 6) reaching 84.4 gS·m^{-3}·h^{-1} (RE = 99%) at an EBRT of 1.6 min. The results obtained with this support material were comparable with results found in the literature for the most common materials used, such as OPUF [18,36] and Pall rings [9], demonstrating that it has potential for larger scale application. As stated previously, using OPUF, Fernández et al. [18] obtained a critical elimination capacity of 130 gS·m^{-3}·h^{-1} at an EBRT of 2.4 min. In the present work, the critical EC was not obtained due to system limitations, but the results were promising considering that PVC has the advantage of being a low-cost material, increasing the biotrickling filter economic viability. The points with different behavior (Figure 6) probably occurred due to H$_2$S inlet load fluctuations, since the biogas supplied in the system was obtained from a real sewage treatment plant from a city with approximately 90 thousand inhabitants. Therefore, the system, in this dimension of operation, was not robust enough to absorb these variations. Additionally, the three points (marked with arrows) with the lowest elimination capacity (two in 2.4 min and one in 1.6 min of EBRT) was affected by the weather which reached 39 °C in the first case and 27 °C in the second case, destabilizing the process due to temperature sensitivity.

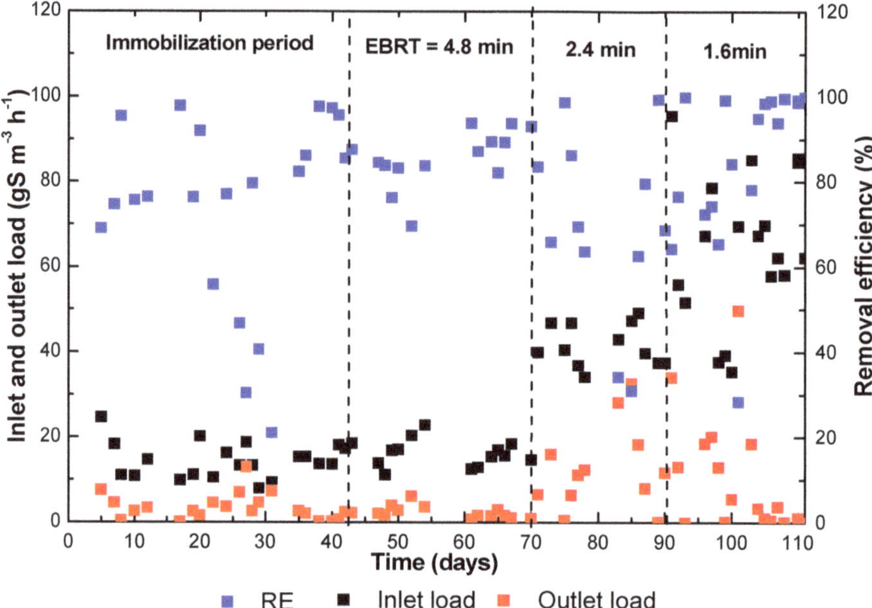

Figure 5. Removal efficiency and inlet and outlet loads during 111 days of biotrickling filter operation.

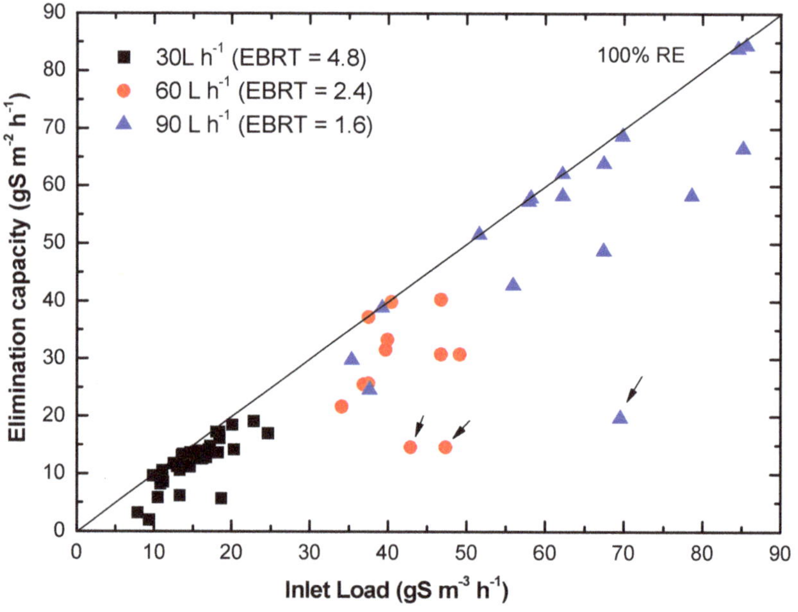

Figure 6. Elimination capacity of the biotrickling filter under different EBRTs.

The temperature effect was analyzed in the first 70 days of operation and as shown in Figure 7a, all ranges of temperatures presented Inlet Loads between 8 and 23 gS·m^{-3}·h^{-1} and also reached ECs higher than 14 gS·m^{-3}·h^{-1} in the operational conditions. It is important to emphasize that despite presenting similar EC behavior for all temperature ranges, the highest RE and lowest variations were obtained for temperatures from 35.5 to 36.7 °C, and temperatures outside this range presented both high distancing from the 100% line and more variations, showing that the system may be outside its optimal condition. This is an important factor for biological H$_2$S removal systems and the optimum obtained was 36 ± 0.7 °C in which the RE observed was 95.72 ± 4.50%. Temperature ranges from 30.0 to 35.2 °C presented REs of 76.31 ± 13.92% and temperatures higher than 36.7 °C demonstrated a RE of 83.50 ± 21.65% (Figure 7b).

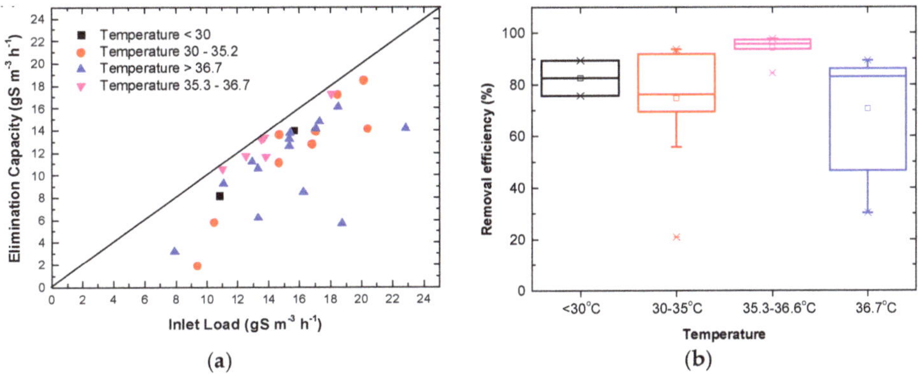

Figure 7. Effect of temperature on the (**a**) elimination capacity and (**b**) removal efficiency. The data include inter-quartile deviations and medians (larger box), (□) average values, (x) outliers and (-) maximum (upper whisker) and minimum (lower whisker) limits of non-discrepant values.

The results obtained using PVC pieces, as an alternative support, presented similar removal efficiency when compared to commercial supports. Table 3 shows a comparison of operation parameters and results from the present work with another study using the same reactor filled with OPUF as support material [19]. It is possible to observe that, for most of the operating time, the biotrickling filter packed with OPUF was operated under low IL and the biotrickling filter with PVC under high IL and despite this, the latter reached similar RE. As a comparison, the system with PVC when submitted to lower IL (12.82 ± 2.47 gS·m^{-3}·h^{-1}) at an EBRT of 1.6 min showed an RE of 87.18%. Other alternative support materials such as "Raschig Ceramic Rings", presented a lower RE (75%) under similar operational conditions [37].

Table 3. Comparison of results from different parameters in a biotrickling filter packed with PVC pieces and OPUF.

Parameter	PVC Pieces Present Study	OPUF [18]
Energy source for immobilization	H$_2$S from biogas	Na$_2$S$_2$O$_3$·5H$_2$O
Nitrate concentration (g·N-NO$_3^-$·L^{-1})	0.25–8.0	0.25–2.0
EBRT (min)	1.6, 2.4 and 4.8	1.6, 2.9 and 6.2
TLV (m·h^{-1})	8–11	4.4, 7.4 and 11
Temperature (°C)	24–40	22–47
H$_2$S Inlet Load (gS·m^{-3}·h^{-1})	8–108	2–16
Elimination Capacity (gS·m^{-3}·h^{-1})	84.4	14
Removal Efficiency (%)	95.72 (IL = 67.38 ± 17.74 gS·m^{-3}·h^{-1}), EBRT = 1.6 min)	98 (IL = 6.13 ± 0.49 gS·m^{-3}·h^{-1}, EBRT = 2.9 min)

The IL applied to the PVC system was from 4 to 6.75 times higher than that applied to the OPUF system, providing conditions to reach 84.4 $gS \cdot m^{-3} \cdot h^{-1}$ of EC, 6 times higher than the EC obtained using OPUF under lower IL. Although under optimal conditions, both PVC and OPUF reached RE values higher than 95%, what is remarkable is that PVC presented this RE even under an IL 137 times higher than OPUF. Both systems, even in the small dimensions studied, demonstrated robustness with high elimination capacity under real industrial conditions, which means seasonal differences in temperature, operation problems, and not controlling the H_2S concentration coming from the treatment plant. It is worth highlighting PVC as an alternative packing material for potential application in larger scales, reducing the initial process costs, which has a high impact considering the amount of packing material required for higher inlet load from real industrial sources

4. Conclusions

The use of biotrickling filtration as a biological technology for biogas biodesulfurization is consolidated as an effective and promising alternative. Nevertheless, the cost savings for the application of this technology, especially in developing countries, is a concern for researchers in order to make it accessible to a greater number of companies. The challenge is to find a packing material with the necessary features for this operation. In a comparative study in columns utilizing four different packing materials, strips of Polyvinyl Chloride (PVC) presented similar efficiency to OPUF, a recognized packing material for decontamination of gases. PVC-biotrickling filter was compared with OPUF- biotrickling filter. Although both had the same dimensions, each filter was assembled in a different place; the PVC-biotrickling filter was coupled to an output of H_2S from a Sewage Treatment Plant (Matão, São Paulo, Brazil) and the other coupled to a wastewater treatment plant of a brewery. Both biotrickling filters presented high elimination capacity; however, the PVC-biotrickling filter was submitted to more aggressive conditions, such as higher IL and temperature variations, maintaining good performance throughout operation time. Thus, PVC can be stated as a potential low-cost packing material for decontamination of industrial gases containing H_2S in a biotrickling filter system.

Author Contributions: D.B. and R.d.B.S.G. conceived and designed the experiments. S.P.T. and L.F.H. performed the experiments. All authors contributed to analyses of data. S.P.T., R.d.B.S.G. and D.B. wrote the paper.

Funding: This research was funded by FAPESP, grant number 2014/22710-2 and 2016/01299-8.

Acknowledgments: The authors wish to thank FAPESP for funding this work through the project 2014/22710-2 and for fellowships to SPT (2016/01299-8) and LFH (2015/24990-5). The authors declare that they not received funds for covering the costs to publish in open access.

Conflicts of Interest: The authors declare no conflict of interest.

References

1. Santos, I.F.S.D.; Barros, R.M.; Tiago Filho, G.L. Electricity generation from biogas of anaerobic wastewater treatment plants in Brazil: An assessment of feasibility and potential. *J. Clean. Prod.* **2016**, *126*, 504–514. [CrossRef]

2. Abatzoglou, N.; Boivin, S. A review of biogas purification processes. *Biofuels Bioprod. Biorefining* **2009**, *3*, 42–71. [CrossRef]

3. Deshusses, M.A.; Cox, H.H.J. A cost benefit approach to reactor sizing and nutrient supply for biotrickling filters for air pollution control. *Environ. Prog.* **1999**, *18*, 188–196. [CrossRef]

4. Schieder, D.; Quicker, P.; Schneider, R.; Winter, H.; Prechtl, S.; Faulstich, M. Microbiological removal of hydrogen sulfide from biogas by means of a separate biofilter system: Experience with technical operation. *Water Sci. Technol.* **2003**, *48*, 209–212. [CrossRef] [PubMed]

5. Jaber, M.B.; Couvert, A.; Amrane, A.; Le Cloirec, P.; Dumont, E. Hydrogen sulfide removal from a biogas mimic by biofiltration under anoxic conditions. *J. Environ. Chem. Eng.* **2017**, *5*, 5617–5623. [CrossRef]

6. Cohen, Y. Biofiltration—The treatment of fluids by microorganisms immobilized into the filter bedding material: a review. *Bioresour. Technol.* **2001**, *77*, 257–274. [CrossRef]

7. Soreanu, G.; Béland, M.; Falletta, P.; Ventresca, B.; Seto, P. Evaluation of different packing media for anoxic H_2S control in biogas. *Environ. Technol.* **2009**, *30*, 1249–1259. [CrossRef]

8. Gabriel, D.; Cox, H.H.J.; Deshusses, M.A. Conversion of Full-Scale Wet Scrubbers to Biotrickling Filters for Control at Publicly Owned Treatment Works. *J. Environ. Eng.* **2004**, *130*, 1110–1117. [CrossRef]

9. Fernández, M.; Ramírez, M.; Pérez, R.M.; Gómez, J.M.; Cantero, D. Hydrogen sulphide removal from biogas by an anoxic biotrickling filter packed with Pall rings. *Chem. Eng. J.* **2013**, *225*, 456–463. [CrossRef]

10. Montebello, A.M.; Bezerra, T.; Rovira, R.; Rago, L.; Lafuente, J.; Gamisans, X.; Campoy, S.; Baeza, M.; Gabriel, D. Operational aspects, pH transition and microbial shifts of a H 2 S desulfurizing biotrickling filter with random packing material. *Chemosphere* **2013**, *93*, 2675–2682. [CrossRef]

11. Qiu, X.; Deshusses, M.A. Performance of a monolith biotrickling filter treating high concentrations of H_2S from mimic biogas and elemental sulfur plugging control using pigging. *Chemosphere* **2017**, *186*, 790–797. [CrossRef] [PubMed]

12. Fortuny, M.; Baeza, J.A.; Gamisans, X.; Casas, C.; Lafuente, J.; Deshusses, M.A.; Gabriel, D. Biological sweetening of energy gases mimics in biotrickling filters. *Chemosphere* **2008**, *71*, 10–17. [CrossRef] [PubMed]

13. Moe, W.M.; Irvine, R.L. Polyurethane foam medium for biofiltration I: Characterization. *J. Environ. Eng.* **2000**, *126*, 826–832. [CrossRef]

14. Lisiecki, J.; Blazejewicz, T.; Klysz, S. Selected properties of auxetic foams made in AFIT. *Tech. News* **2011**, *27*, 42–45.

15. Loy, J.; Heinrich, K.; Egerer, B. Influence of filter material on the elimination rate in a biotrickling filter bed. In Proceedings of the 90th Annual Meeting and Exhibition of the Air and Waste Management Association, Toronto, ON, Canada, 8–13 June 1997; Air and Waste Management Association: Pittsburgh, PA, USA, 1997.

16. Fernández, M.; Rámirez, M.; Pérez, R.M.; Rovira, R.; Gabriel, D.; Gómez, J.M.; Cantero, D. Hydrogen sulfide (H_2S) removal from biogas using biofiltration under anoxic conditions. In Proceedings of the Duke UAM 2010 Conference Biofiltration Air Pollution Control, Washington, DC, USA, 28–29 October 2010; Volume 6.

17. Montebello, A.M.; Fernández, M.; Almenglo, F.; Ramírez, M.; Cantero, D.; Baeza, M.; Gabriel, D. Simultaneous methylmercaptan and hydrogen sulfide removal in the desulfurization of biogas in aerobic and anoxic biotrickling filters. *Chem. Eng. J.* **2012**, *200–202*, 237–246. [CrossRef]

18. Fernández, M.; Ramírez, M.; Gómez, J.M.; Cantero, D. Biogas biodesulfurization in an anoxic biotrickling filter packed with open-pore polyurethane foam. *J. Hazard. Mater.* **2014**, *264*, 529–535. [CrossRef] [PubMed]

19. Guerrero, R.B.; Bevilaqua, D. Biotrickling Filtration of Biogas Produced from the Wastewater Treatment Plant of a Brewery. *J. Environ. Eng.* **2015**, *141*, 04015010-1–04015010-7. [CrossRef]

20. Almenglo, F.; Bezerra, T.; Lafuente, J.; Gabriel, D.; Ramírez, M.; Cantero, D. Effect of gas-liquid flow pattern and microbial diversity analysis of a pilot-scale biotrickling fi lter for anoxic biogas desulfurization. *Chemosphere* **2016**, *157*, 215–223. [CrossRef] [PubMed]

21. Almenglo, F.; Ramírez, M.; Gómez, J.M.; Cantero, D. Operational conditions for start-up and nitrate-feeding in an anoxic biotrickling filtration process at pilot scale. *Chem. Eng. J.* **2016**, *285*, 83–91. [CrossRef]

22. Moran, S. *An Applied Guide to Water and Effluent Treatment Plant Design*; Moran, S., Ed.; Butterworth-Heinemann: Oxford, UK, 2018; ISBN 9780128113103.

23. Tomas, M.; Fortuny, M.; Lao, C.; Gabriel, D.; Lafuente, J.; Gamisans, X. Technical and economical study of a full-scale biotrickling filter for H_2S removal from biogas. *Water Pract. Technol.* **2009**, *4*, 2. [CrossRef]

24. López, L.R.; Bezerra, T.; Mora, M.; Lafuente, J.; Gabriel, D. Influence of trickling liquid velocity and flow pattern in the improvement of oxygen transport in aerobic biotrickling filters for biogas desulfurization. *J Chem. Technol. Biotechnol.* **2016**, *91*, 1031–1039. [CrossRef]

25. López, L.R.; Brito, J.; Mora, M.; Almenglo, F.; Baeza, J.A.; Ramírez, M.; Lafuente, J.; Cantero, D.; Gabriel, D. Feedforward control application in aerobic and anoxic biotrickling filters for H_2S removal from biogas. *J. Chem. Technol. Biotechnol.* **2018**, *93*, 2307–2315. [CrossRef]

26. López, L.R.; Dorado, A.D.; Mora, M.; Gamisans, X.; Lafuente, J.; Gabriel, D. Modeling an aerobic biotrickling filter for biogas desulfurization through a multi-step oxidation mechanism. *Chem. Eng. J.* **2016**, *294*, 447–457. [CrossRef]

27. Montebello, A.M.; Baeza, M.; Lafuente, J.; Gabriel, D. Monitoring and performance of a desulphurizing biotrickling filter with an integrated continuous gas/liquid flow analyser. *Chem. Eng. J.* **2010**, *165*, 500–507. [CrossRef]

28. Fortuny, M.; Gamisans, X.; Deshusses, M.A.; Lafuente, J.; Casas, C.; Gabriel, D. Operational aspects of the desulfurization process of energy gases mimics in biotrickling filters. *Water Res.* **2011**, *45*, 5665–5674. [CrossRef] [PubMed]

29. Vikromvarasiri, N.; Champreda, V.; Boonyawanich, S.; Pisutpaisal, N. Hydrogen sulfide removal from biogas by biotrickling filter inoculated with Halothiobacillus neapolitanus. *Int. J. Hydrog. Energy* **2017**, *42*, 18425–18433. [CrossRef]

30. Gabriel, D.; Deshusses, M.A. Retrofitting existing chemical scrubbers to biotrickling filters for H_2S emission control. *Proc. Natl. Acad. Sci. USA* **2003**, *100*, 6308–6312. [CrossRef] [PubMed]

31. Cano, P.I.; Colón, J.; Ramírez, M.; Lafuente, J.; Gabriel, D.; Cantero, D. Life cycle assessment of different physical-chemical and biological technologies for biogas desulfurization in sewage treatment plants. *J. Clean. Prod.* **2018**, *181*, 663–674. [CrossRef]

32. Hidalgo, L.F.; Santos, J.L.; Tayar, S.P.; Sarti, A.; Palmieri, M.C.; Guerrero, R.B.S.; Bevilaqua, D. Evaluation of substrate consumption kinetics in different support materials for biotrickling filters aiming biogas desulfurization. *Solid State Phenom.* **2017**, *262 SSP*, 682–686. [CrossRef]

33. Solcia, R.B.; Ramírez, M.; Fernández, M.; Cantero, D.; Bevilaqua, D. Hydrogen sulphide removal from air by biotrickling filter using open-pore polyurethane foam as a carrier. *Biochem. Eng. J.* **2014**, *84*, 1–8. [CrossRef]

34. Rodier, J.; Geofray, C.H. *Análisis de Las Aguas. Aguas Naturales, Aguas Residuales, Agua de Mar*, 3rd ed.; Omega: Barcelona, Spain, 1998.

35. APHA/AWWA/WEF. *Standard Methods for Examination of Water and Wastewater*, 21st ed.; American Public Health Association: Washington, DC, USA, 2005.

36. Ramírez, M.; Fernández, M.; Granada, C.; Le Borgne, S.; Gómez, J.M.; Cantero, D. Biofiltration of reduced sulphur compounds and community analysis of sulphur-oxidizing bacteria. *Bioresour. Technol.* **2011**, *102*, 4047–4053. [CrossRef] [PubMed]

37. Ziemiński, K.; Kopycki, W.J. Impact of Different Packing Materials on Hydrogen Sulfide Biooxidation in Biofilters Installed in the Industrial Environment. *Energy Fuels* **2016**, *30*, 9386–9395. [CrossRef]

 *chemengineering*

Review

Desulphurisation of Biogas: A Systematic Qualitative and Economic-Based Quantitative Review of Alternative Strategies

Oseweuba Valentine Okoro * and **Zhifa Sun ***

Department of Physics, University of Otago, P.O. Box 56, Dunedin 9054, New Zealand
* Correspondence: oseweuba.okoro@otago.ac.nz (O.V.O.); zhifa.sun@otago.ac.nz (Z.S.);
  Tel.: +64-3-479-7812 (Z.S.)

Received: 27 June 2019; Accepted: 28 August 2019; Published: 2 September 2019

Abstract: The desulphurisation of biogas for hydrogen sulphide (H_2S) removal constitutes a significant challenge in the area of biogas research. This is because the retention of H_2S in biogas presents negative consequences on human health and equipment durability. The negative impacts are reflective of the potentially fatal and corrosive consequences reported when biogas containing H_2S is inhaled and employed as a boiler biofuel, respectively. Recognising the importance of producing H_2S-free biogas, this paper explores the current state of research in the area of desulphurisation of biogas. In the present paper, physical–chemical, biological, in-situ, and post-biogas desulphurisation strategies were extensively reviewed as the basis for providing a qualitative comparison of the strategies. Additionally, a review of the costing data combined with an analysis of the inherent data uncertainties due underlying estimation assumptions have also been undertaken to provide a basis for quantitative comparison of the desulphurisation strategies. It is anticipated that the combination of the qualitative and quantitative comparison approaches employed in assessing the desulphurisation strategies reviewed in the present paper will aid in future decisions involving the selection of the preferred biogas desulphurisation strategy to satisfy specific economic and performance-related targets.

Keywords: desulphurisation; biogas; in-situ biogas desulphurisation; anaerobic digestion; post-biogas desulphurisation

---

1. Introduction: Hydrogen Sulphide (H_2S) Formation During Anaerobic Digestion and Its Effect on Biogas Utilisation

Bioenergy recovery from biomass resources may be achieved via the employment of several biomass conversion pathways [1,2]. A review of literatures, however, highlights an upsurge in the application of anaerobic digestion (AD) technologies in recent times [3,4]. This observation may be reflective of the strategic preference of this conversion pathway due to its capability to facilitate a simultaneous management of high moisture organic waste streams while also generating bioenergy in the form of biogas at a reduced cost [2,5]. The biogas product is typically composed of carbon dioxide (CO_2) and biomethane (CH_4) which are present in volumetric percentages of 30–40% and 60–70%, respectively [6]. However, although AD technologies for biogas production can be employed in the conversion of a multitude of organic streams to biogas under the action of suitable microbes, the degradation of sulphur containing organics, i.e., proteins, and the reduction of anionic species, i.e., SO_4^{2-}, will result in the associated generation of hydrogen sulphide (H_2S) together with the biomethane [7,8]. The H_2S product may be formed during the anaerobic digestion process via macromolecules undergoing transformations either via an assimilatory pathway or a dissimilatory pathway [9]. These pathways lead to the conversion of sulphur containing compounds (SCC) either to soluble sulphides (HS^-), which is a precursor to H_2S formation, or the consumption of SCC for biosynthesis of amino acids (cysteine and methionine), considered fundamental for intracellular adenosine-5-phosphosulphate activation within the microbes [9,10]. In some cases, however, assimilatory reduction of sulphate ions is achieved via so-called 'trophic reactions', thus leading to direct H_2S formation [9–11]. According to Mackie et al., [9], bacteria capable of assimilatory sulphate reduction are ubiquitous in digestion substrates, with notable examples belonging to the genera of *Veillonella*, *Megasphaera*, and *Enterobacteria*. On the other hand, the transformation of sulphur containing macromolecules via the dissimilatory pathway always results in bisulfide (HS^-) formation rather than the characteristic amino acid production of the assimilatory pathway [9]. Notable examples of microbes capable of sulphur reduction via HS^- formation include neutrophilic *Desulfotomaculum solfataricum* and *Desulfotomaculum thermosapovorans* [12,13]. Microbes of *Desulfotomaculum solfataricum* and *Desulfotomaculum thermosapovorans* are capable of thriving under temperature conditions as high as 60 °C and 50 °C, respectively [12,13]. The formation of HS^- via the dissimilatory pathway occurs via the reduction of inorganic and organic sulphur forms (i.e., sulphates) present in the organic feedstock, which occurs as illustrated by the following example equations [9,14,15];

$$CH_3COO^- + SO_4{}^{2-} + H^+ \rightarrow HS^- + 4H_2O \tag{1}$$

$$C_2H_5COO^- + \frac{3}{4}SO_4{}^{2-} \rightarrow \frac{3}{4}HS^- + CH_3COO^- + HCO_3{}^- + \frac{1}{4}H^+ \tag{2}$$

$$C_2H_5COO^- + \frac{7}{4}SO_4{}^{2-} + \frac{1}{4}H^+ \rightarrow \frac{7}{4}HS^- + 3HCO_3{}^- + \frac{1}{2}H^+ + \frac{1}{4}OH^- \tag{3}$$

$$C_3H_7COO^- + \frac{1}{2}SO_4{}^{2-} \rightarrow \frac{1}{2}HS^- + 2C_2H_5COO^- + \frac{1}{2}H^+ \tag{4}$$

$$C_3H_7COO^- + \frac{5}{2}SO_4{}^{2-} + \frac{1}{2}H_2O \rightarrow \frac{5}{2}HS^- + 4HCO_3{}^- + \frac{3}{4}H^+ + \frac{3}{4}OH^- \tag{5}$$

Equations (1)–(5) illustrate the formation of soluble sulphides (HS^-) via reactions involving intermediate products of the anaerobic digestion (AD) of acetates (Equation (1)), propionates (Equations (2) and (3)), and butyrates (Equations (4) and (5)). This shows that sulphate reducers will outcompete with methanogens because they have better affinities for acetate and H_2 [16]. Examples of microbes capable of dissimilatory sulphur reductions are organotrophic mesophilic isolated strains of *Deltaproteobacteria* with some Archaea (*Archaeoglobus fulgidus*) [17]. It is important to state that dissimilatory sulphate reductions are enzymologically distinct from the assimilatory sulphate reductions. This is because dissimilatory sulphur reductions can only occur under the action of bacteria capable of consuming specific compounds of sulphates, sulphites, or thiosulphates, as opposed to assimulatory sulphur reductions, which are applicable to all sulphur-containing macromolecules [18].

Equations (1)–(5) highlights the formation of soluble sulphides (HS^-) within the digester, which may be converted to dissolved H_2S as follows;

$$H_2S \leftrightarrow H^+ + HS^- \tag{6}$$

where the concentrations of the species of HS^- and H_2S in the aqueous phase are determined using the Henderson–Hasselbach, relationship as follows [19];

$$Log_{10}\frac{[HS^-]}{[H_2S]} = pH + Log_{10}K_a \tag{7}$$

where $[HS^-]$ denotes the concentration of bisulphide ions present in the substrate in molar (M), and $[H_2S]$ denotes the concentration of dissolved H_2S, present in the substrate, in molar (M), *pH* is the pH level of the substrate, K_a denotes the first ionisation constant of H_2S, specified to be 9.1×10^{-8} [20].

Using Equations (6) and (7) above, it can be shown that the concentrations of H_2S and HS^- within the digester are continually in competition, as the pH within the digester changes from 1 to 14, as highlighted in Figure 1.

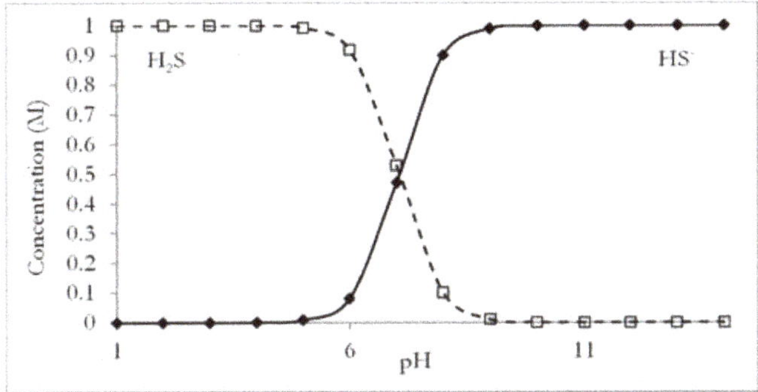

Figure 1. Illustrating the variation of concentrations of dissolved hydrogen sulphide (H_2S) and bisulfide (HS^-) in an aqueous solution with pH of the digester.

The importance of the H_2S content of biogas is reinforced by the possibility of generating biogas with H_2S concentrations of about 18,000 mg/m^3 ($\approx$1.32% v/v) [7,8]. High H_2S concentration in biogas is undesirable, since there is a risk of H_2S release (during biogas collection and transfer) which may lead to pulmonary oedema in humans at concentrations higher than ~400 mg/m^3 ($\approx$ 0.029% v/v) [21]. Furthermore, the utilisation of biogas containing H_2S as a biofuel for combined heat and power generation also has the potential of leading to the corrosion of engines and the rapid degradation of engine lube oil [22]. The combustion of biogas containing H_2S also leads to the formation of sulphur dioxide (SO_2), via the reaction presented in Equation (1) [22,23].

$$2H_2S + 3O_2 \rightarrow 2SO_2 + 2H_2O \tag{8}$$

SO_2 is a precursor for acid rain formation that is responsible for several negative impacts on the environment, such as the destruction of agricultural vegetation and pollution of the surrounding aquatic environment [22,23]. It is also possible for the formation of H_2SO_4, which is a strongly corrosive acid that may be formed when H_2S reacts with water, leading to corrosive effects on equipment [24]. It is therefore necessary to ensure that the H_2S content in biogas is minimised, thus limiting the occurrence of the negative effects discussed above. Indeed, the European committee of standards specifies that the preferred biogas product for use as biofuel should have a H_2S concentration of < 20 mg/m^3 ($\approx$ 0.0015% v/v) [25]. The biogas composition must be ascertained prior to its utilisation, with appropriate strategies employed to limit the H_2S content of biogas and facilitate the generation of 'H₂S-free' biogas.

The present study will therefore provide a qualitative review of major desulphurisation technologies while also incorporating a quantitative dimension to the comparison of these technologies. The importance of this review is reinforced by the absence of publications that provide an exhaustive review of all major technologies integrated with simplified quantitative economic considerations.

2. Methodology Employed

In this present study, an initial qualitative review of the alternative desulphurisation technologies was undertaken via a structured approach which incorporates several steps, namely, relevant study identification, screening of existing study contributions, and the analysis of previously reported data as a basis for establishing logical review conclusions. These steps are identified as crucial requirements in achieving a systematic review of any subject matter [26]. The comparison of alternative desulphurisation technologies was achieved through a rigorous consideration of relevant journal papers and books available from the institute for scientific information (ISI) web of knowledge database. Numerous searches were conducted using key phrases such as 'desulphurisation of biogas' and 'technologies for H_2S removal' to enable an efficient web-search for relevant publications.

After qualitative reviews of the different biogas desulphurisation strategies were concluded, comparative quantitative assessments were undertaken to aid future decisions with respect to the selection of the preferred desulphurisation strategy. Quantitative assessments were undertaken by comparing estimated unit desulphurisation costs. In undertaking a quantitative analysis of the different desulphurisation technologies, difficulties associated with obtaining costing data for the different desulphurisation technologies in a unified manner were appreciated. In other words, it may be difficult to truly represent economic differences for data obtained from different sources. This is a common challenge when comparisons of processes are undertaken based on economic data obtained from the literature, with Ref. [27] employing economic data published in the European Union as an approach to reduce possible interpretational discrepancies. The quantitative review in the present paper therefore sought to utilise data, obtained based on studies undertaken in the member countries of the organisation for economic co-operation and development (OECD). Thus, while a comprehensive unification of the costing data obtained from different countries may be difficult, it is expected that the cost estimates and the conclusions reached may give an overall impression of the most economically favourable desulphurisation technology. In an attempt to further improve the applicability of the costing estimates obtained, the results were assessed via an incorporation of uncertainties for 50% to 150% variations in the annual operating cost and the annualised capital costs. The inclusion of uncertainty considerations is discussed further in Section 2.2 below.

Based on literature obtained costing data, the unit desulphurisation cost, U_c, in US $/m^3$ of biogas was therefore calculated as follows;

$$U_c = \frac{C_{2019} + P_{2019}}{V_{2019}} \tag{9}$$

where C_{2019} and P_{2019} are the annualised capital cost and the annual operating cost in US $, and V_{2019} is the volume of biogas desulphurised in the year of 2019. In the present study it was assumed that all desulphurisation technologies are capable of operating for 7200 h per year. The methods employed in estimating the annualised capital cost and the annual operating cost components are discussed in the subsequent subsections.

2.1. Annual Operating and Annualised Capital Cost per Unit Volume Cost

Depending on data availability, several methods were employed in estimating the annualised capital cost and annual operating cost. In one approach, the capital cost per unit volume of biogas desulphurised, c, in US $/m^3$ and operating cost per unit volume of biogas desulphurised, p, in US $/m^3$ were available in the literature and thus utilised as a basis for estimating the annualised capital cost, C_i, in US $, in the ith year and the annual operating cost, P_i, in US $, in the ith year as follows:

$$cv = C_i \tag{10}$$

$$pv = P_i \tag{11}$$

where v specifies the biogas desulphurisation capacity in m^3.

To reflect possible effects of inflation, the annualised capital cost in the present year of 2019 (C_{2019}) was estimated from the calculated annualised capital cost for the reference year of i (C_i), using the chemical engineering plant cost (*CEPCI*) index as follows [28];

$$C_{2019} = C_i \left[\frac{CEPCI_{2019}}{CEPCI_i} \right] \tag{12}$$

where $CEPCI_{2019}$ was specified as 603.1 [29], which is the mean value reported for the year 2018, as the authors had no access to the $CEPCI$ value for the year 2019 at the time of drafting this manuscript. The value of the $CEPCI_i$ employed will be determined by the reference year, i, from whence the data was obtained.

In another approach, the purchase costs for equipment employed in the desulphurisation technologies were available in the literature. The equipment purchase costs were employed in calculating necessary costing inputs for the estimation of the inside battery costs (ISBL) and the investment cost as follows [30,31];

$$ISBL_{i,vref} = f_L \sum_{a}^{n} Cost_{a,vref} \tag{13}$$

$$T_{i,vref} = 2.05 \times ISBL_{i,vref} \tag{14}$$

where the $Cost_{a,vref}$ denotes the purchase cost of the ath equipment in the desulphurisation system of reference capacity, $vref$ in m^3, $ISBL_{i,vref}$ is the inside battery cost in US $ for the year, i, and for the desulphurisation system of reference capacity, $vref$ in m^3, f_L is the Lang factor specified as 2.47; 2.05 is the conversion factor of $ISBL_{i,vref}$ to the total investment cost, $T_{i,vref}$ in US $ estimated for the year, i, for the desulphurisation system of reference capacity, $vref$ in m^3 [30,31];

The annualised capital cost in US $ for year i, for biogas desulphurisation system, with reference capacity, $vref$ in m^3, and denoted by $C_{i,vref}$ was subsequently estimated as follows [28];

$$C_{i,vref} = T_{i,vref} \left[\frac{(1+j)^n \times i}{(1+j)^n - 1} \right] \tag{15}$$

where n is the plant lifespan assumed to be 10 years and j is the discount rate assumed to be 10%.

The calculation of the annualised capital cost estimate for a different capacity, v in m^3, from the annualised capital cost for the reference gas capacity, $vref$ in m^3, was achieved as follows [32];

$$C_i = C_{i,vref} \left(\frac{v}{vref} \right)^k \tag{16}$$

where C_i is the annualised capital cost for a desulphurisation plant with a capacity of v, in m^3, $C_{i,vref}$ is the annualised capital cost for a plant with a reference capacity $vref$, in m^3, and k is the scaling factor specified as 0.7 [32]. All estimates in Equation (16) are for the year i.

To account for inflation considerations, annualised capital cost in the present year of 2019 (C_{2019}) was estimated using the $CEPCI$, as described in Equation (12). During the course of the review of the economic performances of the alternative desulphurisation technologies, a basis of 1000 m^3/h of H_2S containing biogas (also called sour biogas) has been assumed to be available. It has also been assumed that operating costing estimates obtained from the literature did not change significantly over the years. In addition to the aforementioned assumptions, it is important to recognise that the economic assessments of competing desulphurisation technologies are strongly influenced by peculiar properties (i.e., temperature, H_2S concentration) of the biogas product [33]. This clearly implies that any conclusions based on the economic assessments, are largely indicative, and capable of serving only as a guide to support decisions regarding the applicability of the desulphurisation technologies in practice.

2.2. A Consideration of the Inherent Uncertainties in the Costing Data

Due to the limitations of the economic approach discussed in Section 2.1 above, the applicability of the economic assessment results was tested by assessing uncertainties in the unit desulphurisation cost U_c. The uncertainties in the unit desulphurisation cost U_c were estimated for the different desulphurisation technologies for 50% to 150% variations in the annual operating cost and the annualised capital costs. Investigations into uncertainties in the U_c estimates for the different desulphurisation technologies were achieved via Monte Carlo simulations that employ randomised sequences of estimates to numerically determine its probability density according to the probability density function, defined as follows [28];

$$f(x) = \frac{1}{\sqrt{2\pi} \cdot \sigma} \cdot e^{-\frac{1}{2}\left(\frac{x-\mu}{\sigma}\right)^2} \tag{17}$$

where x represents the values of the U_c for different values of annual operating cost, annualised capital cost, interest rate, and the project lifetime, μ represents the mean value of U_c, and σ represents the standard deviation of U_c.

This approach has been employed in a previous study [28], where the area under the curve of $f(x)$ for the interval (a,b) specifies the probability, β, of U_c existing within that interval (a,b) and is calculated using the probability density function as follows [28];

$$\beta(a < x < b) = \int_{a}^{b} f(x)dx \tag{18}$$

In the present study, 1000 different combinations of the annual operating cost and annualised capital cost estimates have been generated using Minitab ®V17 (Minitab, LEAD Technologies, Inc., Charlotte, NC, USA) to reduce computational time. To aid an easy comparison of the results obtained for the different desulphurisation technologies investigated, a box and whisker plot was utilized, rather than a probability distribution plot, for each individual technology.

3. Qualitative Review of the Strategies for Biogas Desulphurisation

In this section, major desulphurisation technologies have been qualitatively reviewed. The technologies have been categorised into major classifications of physical–chemical desulphurisation strategies and biotechnological desulphurisation strategies. These classifications are discussed in subsequent subsections.

3.1. Physical–Chemical Desulphurisation Methods

Physical–chemical desulphurisation methods typically involve technologies that employ physical or chemical phenomena in preventing or limiting the formation of H_2S during anaerobic digestion processes.

3.1.1. In-Situ Chemical Precipitation

The in-situ chemical precipitation desulphurisation strategy refers to approaches that are localised within the digester and that can limit the conversion of dissolved sulphides in the digester to H_2S via their conversion to insolubles within the digester [34,35]. The formation of insolubles is achieved via the dosing of the digester using chemicals that are capable of converting dissolved sulphides to either insoluble metallic sulphide compounds or insoluble elemental sulphur [35,36]. Generally speaking, divalent (Fe^{2+}) and trivalent (Fe^{3+}) iron salts are the most commonly employed chemicals to enable the precipitation of sulphides [37]. These salts include chlorides of Fe^{3+} and Fe^{2+}, iron (III) oxide-hydroxide (FeOOH), and iron (III) hydroxide (Fe (OH)$_3$) [34,38]. The precipitation of sulphides occurs via reactions as follows [34,38];

$$FeCl_2 + HS^- + H^+ \rightarrow FeS + 2HCl \tag{19}$$

$$FeCl_3 + 2HS^- + H^+ \rightarrow FeS + 3HCl + S \tag{20}$$

$$2\text{FeOOH} + 3\text{HS}^- + 3\text{H}^+ \rightarrow 2\text{FeS} + \text{S} + 4\text{H}_2\text{O} \tag{21}$$

$$2\text{Fe(OH)}_3 + 3\text{HS}^- + 3\text{H}^+ \rightarrow 2\text{FeS} + \text{S} + 6\text{H}_2\text{O} \tag{22}$$

According to SevernWye [39], the chemical dosing desulphurisation approach for sulphite precipitation using salts of Fe^{2+} and Fe^{3+} constitutes an easy approach that can be readily retrofitted to existing plants, such that handling and monitoring concerns are minimal. However, while the method is clearly straightforward, the employment of chemicals to aid sulphide precipitation suggests a constant dosing requirement if the desulphurisation process is to be sustained. The need therefore arises for the introduction of auxiliary equipment, such as pumps, to maintain a chemical supply, thus increasing the number of unit operations [40]. Furthermore, although it is possible to approximate the mass of Fe ions (either Fe^{2+} or Fe^{3+}) employed in dosing the substrate for sulphide precipitation using empirical relations, such as Equation (23) [41], there still exist significant difficulties in controlling the actual degree of desulphurisation that can be achieved [39].

$$Fe = \beta \frac{M_{Fe}}{M_S}\left[\left(\frac{\text{H}_2\text{S}}{f_{\text{H}_2\text{S}}}V_{substrate}\right) + \left(\frac{\Delta\text{H}_2\text{S}}{1000}\rho_{\text{H}_2\text{S}}V_{biogas}\right)\right] \tag{23}$$

where Fe represents the iron ions supplied in g/d, H_2S denotes the total mass of dissolved H_2S in g/m^3, $f_{\text{H}_2\text{S}}$ denotes the mass fraction of total dissolved sulphur present as dissolved H_2S, $\Delta\text{H}_2\text{S}$ denotes the mass of H_2S that is removed from the biogas in ppmv, $V_{Substrate}$ represents feed rate of the substrate in m^3/d, V_{biogas} denotes the biogas flowrate in m^3/d, $\rho_{\text{H}_2\text{S}}$ denotes the density of H_2S in g/L under prevailing conditions of temperature and pressure, M_{Fe} represents the molecular mass of iron in gmole, M_S denotes the molecular mass of sulphur in gmole, and β denotes the correction factor introduced to account for over dosing, since the iron ions may react with other organics, such as phosphates, that may be present in the system. The β value may range from 1.7 to 5 [41].

The challenge of the need for sustained chemical dosing may be responsible for the prevalence of technological applications on small scale operations [42], with a few studies (i.e., the study of [43]) exploring the technology on a large scale. In addition to the need for a sustained supply of chemicals for sulphide oxidation, recent studies have also shown that dosing of substrates with salts of Fe^{2+} and Fe^{3+} may impede the bioavailability of useful phosphates, thus hindering the ability of micro-organisms to metabolise organic substrates for biomethane formation [44,45]. Indeed, in the studies undertaken by Al-Imarah et al. [45] and Ofverstrom et al. [46], volumetric biomethane (and net biogas) reductions of 20% and 30%, respectively, were observed when substrate was dosed with salts of Fe^{2+} and Fe^{3+} compared to the volume of biogas produced in the absence of substrate chemical dosing. Another possible limitation to the utilisation iron salts for sulphide removal is the risk of system clogging due to the accumulation of precipitated metallic sulphides along the piping line [47].

3.1.2. Absorption Technologies

This approach involves the direct interaction of biogas with water or the direct interaction of biogas with suitable organic solvents via conventional gas–liquid contactors (packed bed or spray towers) [48]. Such interactions will enable the physical absorption of H_2S by water or the chemical conversion of H_2S to elemental sulphur or metal sulphide, depending on the organic solvent utilised [48]. The physical absorption of H_2S using water is also called water scrubbing and is based on the physical effect of dissolving gases in liquids, more so, H_2S is significantly more soluble than biomethane under similar conditions [49]. For instance, the solubility of biomethane and the solubility of H_2S at a temperature of 20 °C are reported to be 0.023 g/kg of water and 3.78 g/kg of water, respectively [50]. While the physical absorption desulphurisation approach is a cheap technique that can enable the generation of energy dense biogas characterised with a biomethane content greater than 97% v/v, some issues associated with unwanted microbial growth on the surface of packing material have been reported [51]. These issues of microbial growth may lead to reduced flexibility toward compositional variability of the biogas input and significant water consumption, moreso if the absorption efficiency is to be sustained [51]. As stated briefly above, in addition to the use of water, physical absorption may be achieved using other organic solvents, such as the dimethyl ether of polyethylene glycol (commercially known as Selexol), which has a general formula of $CH_3O(C_2H_4O)nCH_3$ (n = 2 to 9) [48,52]. This dimethyl ether of polyethylene glycol has been reported to show an enhanced affinity for H_2S (five times compared to water) [48,52]. The higher solubility of H_2S in the organic solvent of Selexol suggests that a reduced mass of Selexol will be required compared to the mass of water required for an equivalent absorption efficiency. The utilisation of solvents to enable the absorption of H_2S via its solubility may be loosely referred to as physical absorption. In other cases, chemicals are employed to enhance the removal of H_2S from biogas via the so-called chemical absorption. In other words, the functionality of such chemical absorbers is not solely dependent on the physically solubility of H_2S in the solvent, but also on the oxidation of S^{-2} to S or conversion to sulphide salts via chemical reactions. The use of several chemical reagents as H_2S absorbers have been reported in the literatures, such as NaOH, $FeCl_2$, Fe^{3+}/MgO, Fe (OH)$_3$, Fe^{3+}/CuSO$_4$, and Fe^{3+}/ ethylene diamine tetra-acetic acid (EDTA) [48]. It should be immediately clear that after the initial dissolution of H_2S gas in the chemical reaction solution, oxidation of H_2S in the solution of chemical reagents of $FeCl_2$ and $Fe(OH)_3$ for appropriate iron salt precipitate formation occurs according to the reaction equations previously described in Section 3.1.1 above. According to [53], the utilisation of NaOH in chemical absorption constitutes one of the oldest approaches employed in H_2S removal from biogas, with the associated formation of either Na_2S or NaHS in accordance with the following reaction equations;

$$H_2S + 2NaOH \rightarrow Na_2S + H_2O \tag{24}$$

$$H_2S + NaOH \rightarrow NaHS + H_2O \tag{25}$$

The utilisation of the chemical absorber of Fe^{3+}/EDTA (Fe^{3+} to EDTA ratio ranging from 0.909 to 0.5) enables H_2S removal via its oxidation to elemental sulphur [54]. Since the formation of elemental sulphur is encouraged, the utilisation of chelates to enable H_2S removal is considered a more environmentally favourable approach compared to the utilisation of simple chemical absorbers that produce metallic sulphides [55]. This process of H_2S removal is also known as the oxidative absorption process [56]. EDTA is a chelant that performs the function of preventing the formation of insoluble iron compounds without impeding the ability of iron to undergo reduction during the desulphurisation process and oxidation during the regeneration process as follows [54];

$$H_2S + 2Fe^{3+}/EDTA^{n-} \rightarrow S + 2H^+ + 2Fe^{2+}/EDTA^{n-} \tag{26}$$

with the regeneration of the chelant occurring as follows;

$$O_2 + 4Fe^{2+}/EDTA^{n-} + 2H_2O \rightarrow 4Fe^{3+}/EDTA^{n-} + 4OH^- \tag{27}$$

where 'n' represents the charge of the chelant anion. Oxidative absorption processes using chelates occur at conditions of moderate temperatures of 20–60 °C, atmospheric pressure condition (1 atm), and pH conditions ranging from 4–8 [54]. Most importantly, the capacity to directly regenerate the chelant coupled with an enhanced H_2S (chemical) absorption attainable suggests that lower masses of the chelate will be required, compared to the masses, required when chemicals such as NaOH, discussed earlier above, are employed.

In conclusion, H_2S removal from biogas via absorption process have been demonstrated to present removal efficiencies of 99% and 98% on lab-scale and pilot scale operations, respectively, with a strong argument for its industrial application, since a scaled up absorption process presents a removal efficiency of as high as 97% [57,58].

3.1.3. Adsorption Technologies

The employment of adsorption technologies for the desulphurisation of biogas are sometimes classified based on the nature of the adsorption forces employed in the desulphurisation process [59,60]. Adsorption processes typically occur at conditions of high pressures ranging from 2 to 7 bar and temperatures of about 70 °C [61]. In general terms, adsorption may be classified as either physisorption or chemisorption processes, with their major differences summarised in Table 1.

Table 1. Major differences between physisorption and chemisorption [62,63].

Differentiating Parameter	Physisorption	Chemisorption
Bonding forces	Van der Waals forces bond the adsorbate and absorbent with the adsorbate molecule retaining its gas phase electronic structure.	Stronger covalent forces are employed leading to perturbation of the molecular electronic structure of the adsorbate molecule
Adsorption heat	The enthalpy change during the adsorption process is usually low and ranges from 30 to 40 kJ·mol^{-1}.	The enthalpy change during the adsorption process is usually high and ranges from 80 to 800 kJ·mol^{-1}.
Adsorption layers	Multi-layer adsorption typically occurs.	Monolayer adsorption typically occurs.
Temperature requirement	Varies with the specific adsorbate and adsorbent employed with low temperatures considered as favourable.	Higher temperatures considered favourable.
Kinetics	It is a reversible process.	Irreversible as new compounds are formed at the adsorbent surface.

In the context of the present review interest, the adsorbate is H_2S gas, while the adsorbent may be a solid such as activated carbon, impregnated activated carbon, zeolites, or an iron oxide-based material [64]. Activated carbon is commonly employed as the preferred adsorbent because it is a highly porous material with a high adsorption capacity and is reported as being efficient in the removal of low concentrations of H_2S from biogas via the physisorption process [65]. When an adsorbent such as activated carbon is treated or impregnated using a suitable basic solution, such as NaOH and KOH, the adsorption capacity of the activated carbon is further enhanced, since additional sulphide oxidation processes are occurring via the formation of sulphide salts according to reactions previously described in Section 3.1.1 above. Some studies have employed other chemicals such as $CdSO_4$ for adsorbent 'impregnation' to facilitate H_2S removal as follows [66];

$$CdSO_4 + H_2S \rightarrow CdS + H_2SO_4 \tag{28}$$

Another useful adsorbent is zeolite. Zeolites have been employed in previous studies, with adsorption hypothesised to be achieved via chemisorption in the form of reactions occurring between cations present in the zeolite structure and H_2S as follows;

$$H_2S + C^{n+}(Z) \rightarrow HZ + C^{n+}(HS) \tag{29}$$

where C^{n+} represents the positively charged cations such as K^+, Ca^{2+}, Mg^{2+}, and Na^+ present in the zeolite (Z) structure.

Zeolites have however been reported as presenting a poorer H_2S removal performance than alternative adsorbents, such as activated carbon, under moderate temperatures of 30 °C, although its performance increases significantly after steam treatment [67]. Other studies have demonstrated the possibility of enhancing the H_2S adsorption properties of Zeolites via the addition of atoms of Cu and Zn which can present as d-block metal cations [68]. The improved H_2S adsorption performance of zeolites after steam treatment is hypothesised to be due to the additional chemisorption processes as steam treatment encourages the controlled oxidation of metals, leading to the formation of thin metallic oxides [69]. These additional chemisorption processes are due to the reaction that occurs between the cationic basic oxides and H_2S as follows [69];

$$H_2S + C^{n+}O^{2-} \rightarrow H_2O + C^{n+}S^{2-} \tag{30}$$

Enhanced adsorption of H_2S after atom substitution with Cu^{2+} and Zn^{2+} is largely due to the stronger interaction with H_2S due to the improved stability of resulting CuS and ZnS [68]. Additional effects of increases in the spaces present within the zeolite cages due to the introduction of a divalent metallic ion may further improve the adsorption capacity of zeolites for polar molecules, but present minimal favourable effects on the adsorption of H_2S [68]. The introduction of an appropriate divalent metallic ion may be achieved via ion exchange, pores necking, or prior to an initial adsorption of polar molecules [67]. According to Wang [70], the utilisation of the adsorbent of iron oxide (Fe_2O_3) is one of the oldest techniques employed in the desulphurisation of dry H_2S. Iron oxide adsorbents are capable of H_2S removal via chemisorption reactions that result in the formation of insoluble iron sulphides as follows [70];

$$H_2S + \frac{1}{3}Fe_2O_3 \rightarrow H_2O + \frac{1}{3}Fe_2S_3 \tag{31}$$

It is also possible to employ iron (III) hydroxide as an adsorbent for biogas desulphurisation as follows;

$$3H_2S + 2Fe(OH)_3 \rightarrow 6H_2O + Fe_2S_3 \tag{32}$$

In both cases discussed above, the resulting iron sulphides can be oxidised to enable the formation of elemental sulphur and the regeneration of the iron oxide adsorbent as follows [71];

$$\frac{1}{2}O_2 + \frac{1}{3}Fe_2S_3 \rightarrow S + \frac{1}{3}Fe_2O_3 \tag{33}$$

This oxidation of the resulting iron sulphides would however result in clogging issues in the adsorbent system due to large masses of elemental sulphur produced, causing the need for the intermittent replacement of the adsorption media [70].

According to [72], adsorbents employed in desulphurisation are typically selected based on their adsorption capacity for H_2S, the kinetics of the adsorption process, the durability as reflected by the mechanical properties, stability of chemical properties, and the multi-functionality of removing other contaminants, such as NH_3, from biogas. Additionally, H_2S removal from biogas via adsorption has been reported to be quite effective since H_2S content of gas can be reduced to low residual concentrations of <1 ppm [73]. This desulphurisation strategy is currently employed mainly in small scale operations [74], possibly due to the high specific cost. Furthermore, previous work has also highlighted challenges associated with competitive adsorption between CO_2 and H_2S biogas components, such that the performance of the adsorbent with respect to H_2S removal from biogas is diminished by the presence of CO_2 in biogas [75]. This is due to the CO_2 and H_2S biogas components competing for the same adsorption sites [75].

3.2. Biotechnological Desulphurisation Strategies

These strategies combine innovative technologies and the sulphide oxidative capacities of specific micro-organisms are mainly in-situ microaeration and biofiltration technologies. These technologies are discussed further in subsequent subsections.

3.2.1. In-Situ Microaeration Desulphurisation

As the name implies, it may be intuitively deduced that this technology serves to limit H_2S formation via the incorporation of technologies for aeration of the digester headspace [47,76]. The potential for unwanted H_2S generation during anaerobic digestion processes may be reduced via the oxidation of soluble sulphides in the substrate, with oxidation aided by the presence of sulphur oxidising bacteria, and small volumes of oxygen (or air) [41,77,78]. The introduction of small doses of air at levels of <1 L of O_2 per 1 L of substrate, 1–10 L per 1 L of substrate, and >10 L per 1 L of substrate specify the conditions for micro, limited, and full aeration [42]. Some examples of bacteria capable of encouraging the oxidation of H_2S in the presence of molecular oxygen include *Acidithiobacillus thiooxidans*, *Thiomonas intermedia*, and *Thiomonas perometabolis* [79]. It was suggested in previous studies that microaeration integrated both chemical and biological reactions for H_2S removal [80,81]. Microaeration enables the oxidation of H_2S present in both the gaseous and aqueous phase to elemental sulphur [82]. According to [83], while elemental sulphur is widely reported as the common sulphide oxidation product, the final product of the microaeration desulphurisation process is largely dependent on the prevailing H_2S to oxygen mole ratio. Reactions highlighting the reaction for producing different sulphide oxidation products (S or/and SO_4^{2-}) are as follows [83];

$$H_2S + \frac{1}{2}O_2 \rightarrow S + H_2O \tag{34}$$

$$S + H_2O + \frac{3}{2}O_2 \rightarrow SO_4^{2-} + 2H^+ \tag{35}$$

$$H_2S + 2O_2 \rightarrow SO_4^{2-} + 2H^+ \tag{36}$$

Other studies also highlight the possibility of the oxidation of dissolved sulphide to generate thiosulphates as follows [84,85];

$$2HS^- + 2O_2 \rightarrow S_2O_3^{2-} + H_2O \tag{37}$$

This method is established to be highly efficient, since previous studies have reported the possibility of achieving H_2S removal efficiencies of greater than 99% when operated on a laboratory scale [77]. Long-term large-scale systems have also been reported to present high H_2S removal efficiencies of up to 90%, thus clearly highlighting the functionality of this technology under industrial scenarios [86]. While recognising the high desulphurisation efficiencies attainable via the in-situ microaeration approach, it is possible that oxygen introduction may present negative effects on the methanogenic activity of some microbes [42,87,88]. Additionally, further oxidation of elemental sulphur may lead to the production of sulphates, which may also inhibit methanogenic activity [89]. It must be stated that microaeration has additional drawbacks associated with the risk of clogging of oxygen supply pipes with elemental sulphur from the partial oxidation of soluble substrates [42]. Additionally, sulfuric acid formation due to H_2S oxidation in the presence of water vapour may lead to associated corrosion problems within the digester [90].

In an attempt to overcome the aforementioned challenges associated with in-situ microaeration desulphurisation strategy, recent investigations have explored the possibility of introducing a membrane to separate the biogas from the oxygen supply pipes [91,92]. In the studies presented in Ref. [91] and Ref. [92], it was demonstrated that it was possible to utilise an internal silicone membrane and a polyvinylidene fluoride membrane to prevent the clogging problems while maintaining high sulphide removal efficiency of 96% and 99.99%, respectively. These challenges may also be limited by the utilisation of a compartment, distinct from the anaerobic digestion vessel, where the microaeration process for H_2S oxidation is taking place [93].

3.2.2. Biofiltration Technologies

Biofiltration technologies employed in the removal of H_2S from biogas include biotrickling filters (BTFs), biofilters (BFs), and bioscrubbers (BS). Biotrickling filters (BTFs) are bioreactors that employ chemically inert packing materials used in immobilising micro-organisms that are capable of sulphide oxidation while also localising wet organic materials as a nutrient source for microorganisms, thus forming a so-called 'filter bed' [94,95]. These microbes grow as biofilms on the packing materials within the filter bed [96]. Given that biofilm formation occurs on the packing material, the packing material that best supports microbial growth must be selected. Typically, the best packing materials are materials that have large surface areas relative to their bulk volumes, such as plastic fibres, honeycomb, or open pore polyurethane foam, which are characterised by specific surface areas greater than 600 m^{-1} [97–99]. Larger surface areas are important since it encourages microbial growth [100]. Other useful properties of packing materials include hydrophobicity and its resistance to degradation from mechanical, chemical, or biological attacks [99]. Additionally, it is important that the moisture content of the filter bed is maintained either by spraying the filter bed with water or by ensuring that a high enough humid gas flow is maintained [101]. BTFs are particularly effective in the removal of H$_2$S, since they support diverse bacteria populations [102]. This may be the basis of its ability to remove other contaminants such as NH$_3$, volatile organic compounds (VOCs), and odours with great efficiency [103,104].

The operation of BTFs involves the passing of the contaminated biogas through the wet filter bed over which an aqueous phase is continuously trickled [95]. The passage of the contaminated biogas thorough the wet filter bed enables the dissolution of H$_2$S gas contaminant to the aqueous phase, followed by the diffusion of the dissolved H$_2$S to the biofilm containing the microbes [105]. The dissolved H$_2$S is subjected to microbial action, leading to its oxidation to elemental sulphur or sulphates [105]. This BTF technology may be considered as a mature technology which has been shown to present high H$_2$S removal efficiencies of ≥99% when employed on bench, pilot, and full-scale operations [106–108].

It is important to note that there is a simpler configuration of the BTF called the biofilter [109]. Simple block diagrams to highlight the difference between the BTF and the simpler biofilter are presented in Figure 2. Figure 2 clearly shows that a biofilter [B] differs from the BTF [A] in terms of the movement of nutrient solution, since the nutrient solution is only occasionally replenished compared to the BTFs, where there is a sustained tricking of nutrient laden water for enhanced nutrient circulation within the filter bed [110]. Generally speaking, biofilters constitute an interesting technology that has been explored in previous studies as bench scale [111] and pilot scale operations [112], with large-scale biofilters expected to lead to further improvements in the economic performance [41].

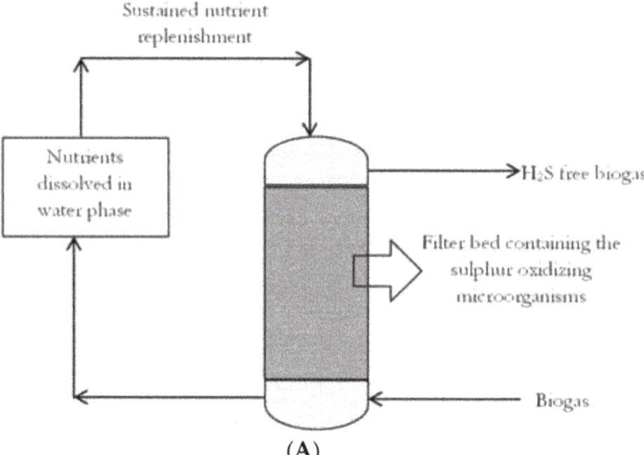

Figure 2. *Cont.*

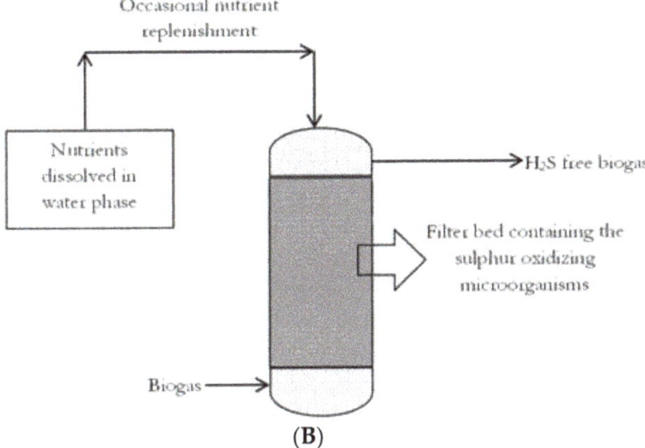

Figure 2. Simple illustration of a biotrickling filter (**A**) and a biofilter (**B**) desulphurisation set for biogas.

Although Figure 2 highlights the configurational simplicity of the biofilter compared to the BTF, biofilters have the major drawback of presenting poor performances, largely due to poor microbial density in the filter beds [109]. Poor microbial density may be a result of reduced access to nutrients by the microbes. Issues of filter bed clogging are, however, common in biofilters and BTFs, suggesting the need for regular cleaning operations [109]. In an attempt to mitigate the aforementioned limitations, biofilter technologies have been developed that exhibit properties similar to biotrickling filters such that water nutrient solution addition, although not continuously available, is supplied intermittently at a higher frequency [113]. There is also some interest in the possibility of utilising sulphur oxidising fungi rather than sulphur oxidising bacteria, since fungi present a higher resistance to acids and low moisture conditions, compared to bacterial biofilms [113]. Bioscrubbers constitute another useful technology that has been shown to be quite effective when employed in pilot scale operations [114]. However, while there are currently a significant number of installations of large-scale biotrickling filters and biofilters globally, the installations of large-scale bioscrubbers constitute the dominant biofilteration technology in small niche markets [113]. The bioscrubbers are able to facilitate the separation of the unwanted H_2S present in the biogas via the incorporation of an initial absorption stage for the dissolution of H_2S gas, followed by the biological treatment of the resulting H_2S containing solution using a reactor containing a suitable microbial population [115,116]. The incorporation of these two stages of H_2S absorption and biological sulphide oxidation is highlighted in Figure 3.

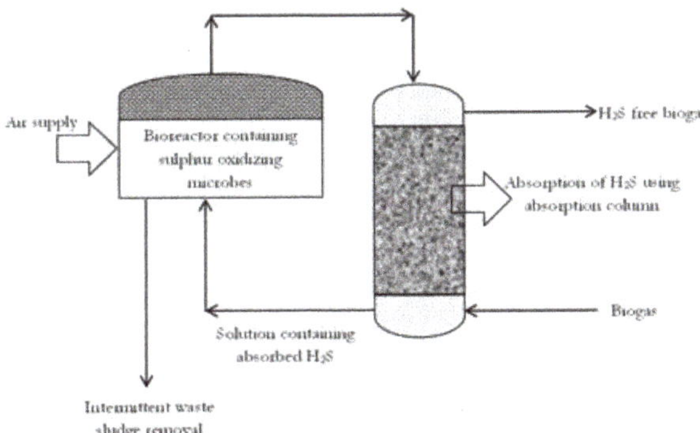

Figure 3. Simple illustration of a bioscrubber highlighting the two steps of initial H_2S absorption followed by H_2S oxidation.

The absorption of H_2S biogas can be achieved using a suitable solvent (i.e., alkaline solution) and technologies such as a spray empty column, a packed column, or a venturi scrubber [117]. The oxidation of the dissolved sulphide is achieved under the action of relevant sulphide oxidising bacteria that oxidises H_2S to either produce S and/or SO_4^{2-} [117]. General speaking, autotropic sulphur oxidising bacteria are typically photoautotrophs or chemolithotrophs [118], and may be employed in BTF, BF, and BS technologies. Due to the importance of photoautotropes, further discussions regarding these sulphur oxidising bacteria have been presented in a stand-alone section below.

Chemolithotrophs, on the other hand, are able to oxidise the sulphides using oxygen (i.e., aerobic species) as electron acceptors and as presented in microaeration discussions above, or nitrate and nitrite (anaerobic species) as electron acceptors. Discussions with respect to the sulphur oxidising bacteria capable of utilising oxygen as electron acceptors have already been discussed in the in-situ microaeration section above, thus will not be discussed further to avoid repetition.

Sulphide oxidising bacteria that employ nitrates (NO_3^-) or nitrites (NO_2^-) present in the filter bed and absorption column of the BTFs or BFs and BS, respectively, as electron acceptors belong to a family of colourless sulphide-oxidising bacteria [119]. These microbes are sometimes referred to as autotropic denitrifies [119] and may include morphologically distinct microbes such as *Thiobacillus denitrificans* [120] and *Thiomicrospira denitrificans* [121]. The autotropic denitrifies are present in highly diverse ecosystems such as soils, sediments, and hydrothermal vents [122]. The oxidation of dissolved sulphides using nitrates or nitrites as electron acceptors may lead to the formation of elemental sulphur or sulphate and nitrite or nitrogen, depending on the sulphide to nitrates or sulphide to nitrites mole ratio, as illustrated in reaction equations as follows [119,123];

$$S^{2-} + 1.6NO_3^- + 1.6H^+ \rightarrow SO_4^{2-} + 0.8N_2 + 0.8H_2O \tag{38}$$

$$S^{2-} + 0.4NO_3^- + 2.4H^+ \rightarrow S + 0.2N_2 + 1.2H_2O \tag{39}$$

$$S^{2-} + NO_3^- + 2H^+ \rightarrow S + NO_2^- + H_2O \tag{40}$$

$$S^{2-} + 4NO_3^- \rightarrow SO_4^{2-} + 4NO_2^- \tag{41}$$

$$HS^- + 0.25NO_3^- + 1.5H^+ \rightarrow S + 0.25NH_4^+ + 0.75H_2O \tag{42}$$

Further oxidation of elemental sulphur and thiosulphides may also occur in the presence of nitrates to produce sulphates as follows [119];

$$S + 1.2NO_3^- + 0.4H_2O \rightarrow SO_4^{2-} + 0.6N_2 + 0.8H^+ \tag{43}$$

$$S_2O_3^{2-} + 1.6NO_3^- + 0.2H_2O \rightarrow 2SO_4^{2-} + 0.8N_2 + 0.4H^+ \tag{44}$$

A notable study that employed autotropic denitrifies in achieving the desulphurisation of biogas is presented in Ref. [97] where BTFs employed nitrate-rich effluent of sequencing batch reactor as both the bacteria source and the nutrient solution. The SBR effluent was also used as the source of microorganisms for the inoculation. Another study presented in Ref. [124] also investigated H_2S removal using an anoxic BTF, which was operated using synthetic wastewater as the nitrate source. Another study undertaken in Ref. [125] not only tested the effectiveness of H_2S removal in an anoxic biotrickling filter, but also demonstrated that the composition of the microbial community varied for different H_2S and NO_3^- loads. Generally speaking, the employment of biological agents in facilitating the oxidation of sulphides as in the case with in-situ microaeration, BTF and BF can intuitively be considered a more environmentally friendly approach compared to the in-situ chemical dosing strategy discussed above, since all inputs occur in nature with no chemicals employed. It is the view of the authors of the present paper that the major issues associated with employing biological desulphurisation strategies may include the problems of microbe recovery, microbial culturing, and sustained microbial growth via temperature and pH control. Bioscrubbers have the advantages of having improved predictability, reliability, and pH control, with reduced clogging issues compared to the BTFs and biofilters [116]. Bioscrubbers are, however, characterised by lower abatement efficiencies, a more complicated start up, and waste sludge disposal issues [116,126]. The biotechnological approaches discussed so far have been demonstrated to be a sufficient and competitive treatment technology for biogas conditioning when compared to physical–chemical technologies [94,95,127].

3.2.3. Phototrophic Sulphur Oxidation

In addition to the chemolithotrophic sulphur oxidising bacteria discussed in Section 3.2.2 above, some previous studies have also demonstrated the possibility of employing phototrophic bacteria to oxidise sulphides to elemental sulphur and sulphates [128,129]. These phototrophic bacteria may be incorporated in the biotechnologies discussed above. The bacteria have the capability to oxidise sulphides to elemental sulphur and sulphate in the presence of light, CO_2, and necessary nutrients, with the intensity of the light required dependent on the concentration of sulphides to be oxidised [129]. The relationship between the concentration of the H_2S oxidised and light intensity is aptly highlighted using the van Niel curve as presented in Figure 4.

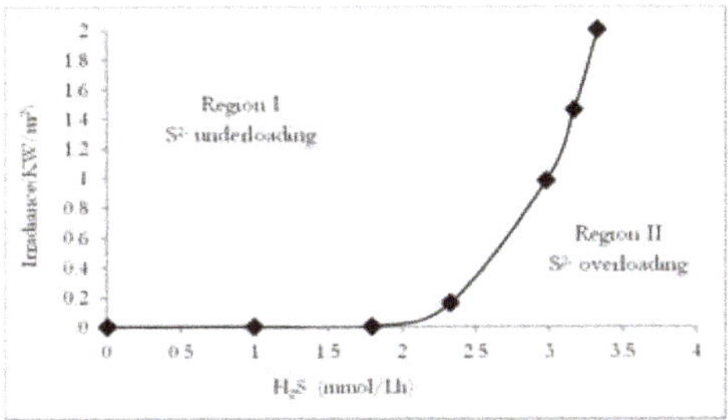

Figure 4. The van Niel curve (adapted from [130]).

Biological desulphurisation via phototropic bacteria activity for H_2S removal occurs according to the general equation below [128,129];

$$2H_2S + CO_2 \xrightarrow{\text{Light}} 2S + CH_2O + H_2O \tag{45}$$

Under conditions of constant high light intensity and low sulphide concentration, previous studies suggested that biological desulphurisation may lead to the formation of sulphates as follows [128,129];

$$H_2S + 2CO_2 + 2H_2O \xrightarrow{\text{Light}} SO_4^{2-} + 2CH_2O + H_2O \tag{46}$$

In Equations (45) and (46), CO_2 functions as the electron acceptor (oxidising agent). According to Pokorna and Zabranska [131], commonly used phototrophic bacteria employed in desulphurisation include the sulphide-oxidizing green bacteria (genera of *Chlorobium* and *Chromatium*) and the sulphide-oxidizing purple bacteria (genera of *Allochromatium, Chromatium, Thioalkalicoccus*). It is important to reiterate that although this desulphurisation strategy achieves the oxidation of dissolved sulphides in the absence of oxygen, some of these microbes have the capability for biological desulphurisation even when molecular oxygen is present [132]. While recognising that there are a multitude of bacteria capable of oxidising dissolved sulphides, such that each bacterial genera is characterised by unique physiological and biological characteristics, the present review will provide a brief summary of some commonalities and differences between different sulphur oxidising bacteria. A brief summary of some characteristics of sulphur oxidising bacteria has been provided, since an extensive investigation of sulphur oxidising physiology bacteria does not constitute an objective of this review paper. Some notable characteristics of some sulphur oxidising bacteria are therefore summarised in Table 2.

Table 2. Summary of some notable properties of sulphur oxidising bacteria considered in the present study.

Some Properties	Sulphur Oxidising Bacteria			
	Photoautotrophs		Chemolithotrophs [a]	References
	Green Sulphur Bacteria	Purple Sulphur Bacteria		
Sulphide oxidation pathway	Reduces carbon dioxide to carbohydrates via the H_2S oxidation.	Also capable of undergoing the phototropic conversion of carbon dioxide to carbohydrates via H_2S oxidation.	May oxidize inorganic sulphur compounds using oxygen, generating energy (aerobic species such as *Beggiatoa* sp) May also oxidize inorganic sulphur compounds using nitrogen oxides for energy generation. Also called autotropic dentrifiers (i.e., anaerobic species such as *Thiomicrospira denitrificans*.	[133,134]
Physiology	Green bacteria are either non-motile green or gliding filamentous green bacteria. Gas vesicles are present and responsible for enhanced buoyancy. Chlorosome complexes are also present and that serve as 'photosynthetic antennas'.	Most purple bacteria are flagellated. Typically, gas vesicles and chlorosomes are not present.	The chemolithotrophs are typically considered colourless due to the absence of photopigments. They may be motile, filamentous organisms (i.e., *Thiobacillus denitrificans, Beggiatoa*) as the mobility assists in migration to regions of higher oxygen concentration. They may also exist as non-motile organisms (i.e., *Thiomicrospira denitrificans*)	[119,135–139]
Light response	Green sulphur bacteria are phototrophs thus require light. They absorb longer wavelengths of light than purple sulphur bacteria since have they have special adaptations to low light. A notable example is the green bacteria is *Chlorobium phaeobacteroides* which is reported to be capable of surviving under a light level of <0.25 μmol photons $m^{-2}\ s^{-1}$.	Purple sulphur bacteria are phototrophs thus also require light. They require shorter wavelengths of light, highlighting the need for higher energy requirement by the purple bacteria.	Photo-inhibition even in low light intensities has been reported in previous studies, with protection from light considered desirable.	[133,140,141]

Table 2. *Cont.*

Some Properties	Sulphur Oxidising Bacteria			
	Photoautotrophs		Chemolithotrophs [a]	References
	Green Sulphur Bacteria	Purple Sulphur Bacteria		
Oxygen tolerance	Strict obligate anaerobes	Can survive in the presence of molecular oxygen (facultative)	Some are dependent on oxygen, thus they are often localised at the interface between aerobic and anaerobic zones where low concentrations of oxygen and sulphide can coexist (i.e., *Beggiatoa* sp and *Thioploca* sp are aerobic). Other colorless bacteria may exist as absolute obligates (i.e., *Thiomicrospira denitrificans*) with some species (i.e., *Thiobacillus denitrificans*) capable of consuming low oxygen concentrations.	[132,139,142,143]
Sulphur availability	Elemental sulphur is deposited extracellularly, free of any encapsulation by proteins. If the sulphides are depleted in the substrate, extracellular sulphur globules may be oxidised completely to sulphate.	Elemental Sulphur is typically deposited intracellularly as spherical particles and is encapsulated by proteins. The sulphur is further oxidised and released from the cells as sulphates. Some species of the purple sulphur bacteria such as *Ectothiorhodospira*, *Halorhodospira*, *Thiorhodospira* also have the capability to deposit elemental sulphur extracellularly.	There may be an intracellular accumulation of elemental sulphur. In some cases, up to 30% of dry cell mass is sulphur (i.e., *Beggiatoa* sp). Large internal vacuoles facilitated the storage of sulphur within the cell wall	[131,144–147]
Sulphide tolerance	Green sulphur bacteria exhibit a high tolerance for high concentrations of sulphides in solution of up to 5–10 mM. This is largely because sulphide oxidation occurs extracellularly.	High sulphide concentrations (5–10 mM) are considered toxic to purple sulphur bacteria largely because sulphides are oxidised internally.	They do not tolerate very high sulphide concentrations (i.e.,~>1920 mg/L).	[146,148–150]
CO_2 fixation pathway	CO_2 fixation is typically achieved via the reductive tricarboxylic acid (RTCA) cycle.	The reductive pentose phosphate (also called Calvin-Benson- Bassham) cycle is typically employed in CO_2 fixation.	CO_2 fixation is typically achieved via the reductive ribulosediphosphate (Calvin) cycle. Some may also be able to utilise small amounts of exogenous organic carbon as carbon sources.	[143,151–153]
Bacteriochlorophyll forms of types a,b,c,d,e and g (letters to illustrate slight changes in structure)	Bacteriochlorophylls of c, d, and e are present	Bacteriochlorophylls of a and b are present	Photopigments are absent.	[154]

[a] The characteristics of aerobic and anaerobic chemolithotrophs tend to overlap.

4. Quantitative Analysis of the Desulphurisation Alternatives

A review of literatures provides some costing input that would be valuable in undertaking economic assessments and uncertainty analysis of the alternative desulphurisation technologies. The literature obtained costing input are summarised in Tables 3 and 4. Since the data presented in Tables 3 and 4 are obtained for the year 2015 and 2011, the associated $CEPCI_{ref}$(s), discussed earlier in Section 2 above, are specified as 556.8 [29] and 585.7 [155], respectively.

Table 3. Literature obtained equipment purchase cost and operating cost for desulphurisation processes of in-situ chemical dosing and in-situ microaeration desulphurisation approach [32].

Desulphurisation Method [a]	Total Operating Cost (US $/y) [b]	Equipment Purchase Cost (US $) [b]
In-situ chemical dosing	12,684.672	0 [d]
In-situ microaeration	3850.704 [c]	25,184.7 [c]

[a] 157.3 m^3/h of biogas; [b] Cost data based on estimates for the year 2015 with currency conversion of 1€ to US $1.12; [c] Mean costing data values for oxygen concentrations of 100%, 95%, and ~21% v/v; [d] The chemicals were specified to be supplied together with the substrate feed at a rate of 1.8 mole Fe per mole of S removed, thus additional supply pathways (separate pipes) were deemed unnecessary.

Table 4. Literature obtained economic data for desulphurisation processes of chelating iron scrubbing, bioscrubbers, biofilter, biotrickling filters, absorption, and adsorption technologies [41].

Desulphurisation Method	Capital Cost per Unit Gas Volume (US $ per m^3) [a]	Operating Cost per Unit Gas Volume (US $ per m^3) [a]
Chelating iron [b]	0.170	0.070
Bioscrubbers	0.160	0.020
Biofilter	0.090	0.030
Biotrickling filters [c]	1.480	0.010
Absorber [d]	2.334	0.018
Adsorption	1.200	0.009

[a] Absorber using chelation iron; [b] Cost data based on estimates for the year 2011 with currency conversion of 1€ to US $1.12; [c] Mean data for biotrickling filters; [d] Mean data for simple absorbers using NaOH as absorbing solution.

The obtained unit cost (U_c) incurred in desulphurising a unit volume of biogas for the competing desulphurisation technologies discussed earlier above have been determined using methods described in Section 2, with results summarised in Table 5. In the present study, distinctions have not been made between the bacterial forms (denitrifying i.e., *Thiobacillus denitrificans*, *Thiomicrospira denitrificans*, or phototrophic) identified as capable of facilitating the biological oxidation of dissolved sulphides [120–122] using biotechnologies of BTFs, BFs, and BS, as discussed earlier above. Also, a generic adsorption desulphurisation process has been specified to simplify the analysis. Given that the purity of the oxygen source for desulphurisation via in-situ micro aeration may vary in terms of the purity of the oxygen supplied (i.e., ranging from 100% oxygen to air contain oxygen), the mean values of costing data for the in-situ microaeration desulphurisation technology utilising 100% of O_2 (v/v), 95% of O_2 (v/v) and air (~21% v/v) [32] as electron donors have been employed as the representative data for simplicity. Employing the literature obtained economic data presented in Tables 3 and 4 the estimated annualised capital cost, annual operating cost, and the unit desulphurisation cost for the competing desulphurisation technologies are presented in Tables 5 and 6.

Table 5. Estimated ISBL cost, investment cost, annualised capital cost, and total operating cost for desulphurisation processes of in-situ chemical dosing and microaeration.

Desulphurisation Method	Annual Operating Cost (US $) [a]	ISBL Cost-Reference Capacity (US $) [b]	Investment Cost, for Reference Biogas Capacity in Year 2015 (US $) [c]	Investment Cost for New Biogas Capacity for Year 2015 (US $) [d]	Annualised Capital Cost for Year 2019 (US $)	Unit Cleaning Cost (US $/m^3-biogas)
In-situ chemical dosing	72,000	0	0	0	0	0.0100
In-situ microaeration	24,488	69,670.95	142,825.46	521,309.35	9189.517	0.0161

[a] Operating cost estimate per year for desulphurisation of biogas at capacity of 1000 m^3/h for a system operating for 7200 h/y; [b] Inside battery limit (ISBL) cost estimate for desulphurisation of biogas at feed rate of 157.3 m^3/h, year 2015 assuming a year operation duration of 7200 h; [c] Total investment cost estimate for desulphurisation of biogas at feed rate of 157.3 m^3/h, year 2015 assuming a year operation duration of 7200 h; [d] Total investment cost estimate for desulphurisation of biogas at feed rate of 1000 m^3/h, year 2015 assuming a year operation duration of 7200 h.

Table 6. Estimated annualised capital cost, annual operating cost and unit desulphurisation cost for desulphurisation technologies of the use of chelating irons, bioscrubbers, biofilters, biotrickling filters, adsorption, and absorption methods.

Desulphurisation Method	Annualised Capital Cost (US $) [a]	Annual Operating Cost [a]	Unit Desulphurisation Cost (US $/m^3)
Chelating iron	1,260,362.64	504,000	0.245
Bioscrubbers	1,186,223.66	144,000	0.185
Biofilter	667,250.81	216,000	0.123
Biotrickling filters	10,972,568.89	72,000	1.534
Absorption	17,304,037.70	129,600	2.421
Adsorption	8,896,677.48	64,800	1.245

[a] Estimated for year 2019.

It is immediately clear from Tables 5 and 6 that biogas desulphurisation via in-situ oxidation of dissolved sulphides via in-situ chemical dosing, for the base case scenario, constitutes the cheapest desulphurisation strategy. However, recognising the risk of variations in the estimated unit desulphurisation cost due to uncertainties in the underlying assumptions incorporated in estimating the annualised capital and operation cost, Monte Carlo simulations have been undertaken, as discussed in Section 2 above. The results are presented in Figure 5.

Figure 5 shows the unit costs for different desulphurisation technologies investigated in the present study. It shows that the unit desulphurisation cost associated with in-situ chemical dosing (CD) and microaeration (MA) approaches are estimated to be the cheapest and second cheapest approaches, with their estimated median unit desulphurisation costs of US $0.01/m^3 and US $0.015/m^3, the unit costs of which range between US $0.003/m^3 and US $0.004/m^3, respectively. The low unit desulphurisation of CD is consistent with a previous study where reduced capital, maintenance, and repair costs were identified as significant advantages of the CD approach compared to other 'post-digestion' biogas scrubbing desulphurisation approaches [156]. It must however be emphasised that, as stated earlier, the introduction of chemicals into the digester to minimise H_2S formation may also present negative effects on microbial activity, leading to diminishing biogas yields. In addition to possible challenges of diminished microbial activities due to the introduction of chemicals to the digester, there may be additional environmental costs associated with the disposal of the resulting sludge containing Fe^{2+} and Fe^{3+} compounds in surrounding land. This is because although iron (i.e., Fe^{2+} and Fe^{3+}) constitutes an essential nutrient for plants, excessive amounts in soil may have toxic effects on plants due to its accumulation within plant cells [157]. The higher cost of the MA desulphurisation method compared to the CD method may be due to the additional cost associated with the purchase of auxiliary technologies, such as pumps for oxygen and air supply. Additionally, some studies have suggested the possibility of negative effects on biogas yields, due to the deterioration of the methanogenic activity of obligatory microbes necessary for biomethane production [42,87,88]. The possibility of these negative effects was however contradicted in studies presented in [158–160]. These contradictory observations suggest that the negative effect on biogas yields may depend on several factors, such as the level of oxygen tolerance by the methanogens present and the nature of the sludge substrate (i.e., granular or flocculent) [42]. This is because the sludge type may also influence the level of exposure of the methanogens to oxygen [42].

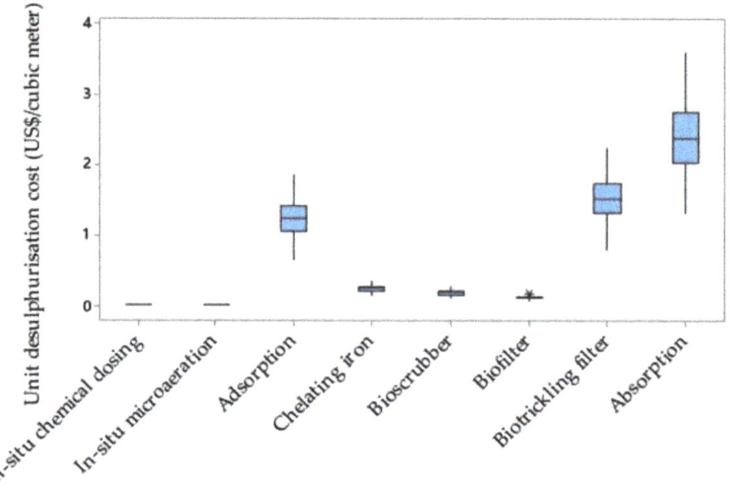

Figure 5. The unit desulphurisation cost distribution for technologies investigated in this paper. The mean unit desulphurisation cost of the desulphurisation technologies were investigated and their associated standard deviation are also presented. These values have been estimated for 50% to 150% changes in the estimated operating cost and capital cost components.

Figure 5 also shows that the traditional absorption desulphurisation strategy using NaOH constitutes the most expensive desulphurisation strategy, with a median unit cost of US $2.38/m^3, which is about 10 times the cost of H_2S absorption using the chelating iron, which presents an estimated median unit cost of US $0.244/m^3. This high cost may explain why this approach is, at present, rarely employed as a viable desulphurisation strategy [53]. In addition to the high cost of traditional absorbers using NaOH, a high technical requirement is necessary to handle the caustic (NaOH) solution and manage associated secondary pollution challenges, since NaOH is typically not regenerated from its Na_2S and NaHS reaction products. The unit desulphurisation cost of the adsorption system is estimated as the third most expensive desulphurisation strategy with a median cost of US $1.231/m^3. The high unit desulphurisation cost estimated may be due to the cost incurred from the regular replacement of adsorbents as a result of clogging issues. This unit cost may be the reason for its preferred utilisation in desulphurisation operations when the concentration of the H_2S in the biogas is relatively low [73]. Interestingly, Figure 5 also shows significant overlap the unit desulphurisation costs of the adsorption and biotrickling filter technologies. This overlap is emphasised when it is observed that the highest possible cost (the whiskers) using the adsorption system of US $1.842/m^3 exceeds the median cost using the adsorption system of US $1.52/m^3. This observation may suggest a close competition between the desulphurisation strategies of adsorption and biotrickling filter technologies based on strictly economic measures.

A comparison of the unit desulphurisation costs using biotrickling filters (BTFs), biofilters (BFs) and bioscrubbers (BS) shows that the BF system is the cheapest desulphurisation strategy. This observation is not unexpected since BF presents a simple design, composed of a fixed bed, equipped with cheap bedding materials and easily acquired microbes. On the other hand, the high median unit cost estimated as US $1.52/m^3$ for desulphurisation operations using biotrickling filters may be reflective of the additional cost required to sustain nutrient supply via the use of auxiliary trickling systems.

Noteworthy Considerations

As a note of caution, it is crucial to recognise that although the present study has presented the in-situ chemical dosing desulphurisation strategy as the most economically favorable approach, as illustrated by its low median unit desulphurisation cost of US $0.01/m^3$, several factors may impede its global acceptance. Some of these factors may include challenges of maintaining a supply of chemicals to the digester if the H_2S oxidation process is to be sustained, possible negative effects on the biogas potential, and the risk of potentially negative environmental outcomes from the disposal of the sludge residues containing insoluble metallic precipitates (i.e., FeS). Logic may however suggest that this desulphurisation approach will be preferred when a small-scale desulphurisation operation requiring high rates of production of H_2S-free biogas is necessary. In such a small-scale operation, the associated chemical cost and unfavourable environmental implications may not be considered significant enough to discourage its application. In other cases, the peculiar properties of the biogas being desulphurised may influence the selection of the appropriate technology to be employed. For instance, for cases involving the desulphurisation of biogas containing low concentrations of H_2S, it may be more practical to implement the desulphurisation process using adsorbents. Also, in the absence of sufficient headspace in the digester, it is not uncommon for the utilisation of biotrickling filters to be preferred to the utilisation of in-situ microaeration, even though biotrickling filters constitute a more expensive approach [15].

Beyond economics, performance, and environmental impact considerations, another factor that may be considered in the selection of the appropriate desulphurisation technology to be employed may be their life cycle impacts, as investigated in the study of [161]. Clearly, therefore, the selection of the appropriate desulphurisation strategy may depend on not only a consideration of the economic performance of the different strategies, but also their respective capabilities to satisfy peculiar design concerns, such as environmental and desulphurisation efficiency performances. It may therefore be prudent to incorporate some form of decision support system, as proposed in Ref. [162], more so if the comparative economic performances of the desulphurisation technologies do not constitute the only design selection criterion.

In another approach, the integration of several desulphurisation strategies may be preferable to the use of a single desulphurisation approach, since such an integration will facilitate the incorporation of strengths while limiting the weaknesses of the component desulphurisation strategies of the different strategies. For instance, the utilisation of an iron sponge composed of Fe_2O_3 for chemical oxidation of HS^- and inoculated with thiobacteria for biological oxidation of H_2S may serve to reduce the mass of metallic precipitates generated, while also ensuring that the desulphurisation process is achieved in a short time.

5. Conclusions

Recognising the negative implications of high H_2S content in biogas on human health, environmental outcomes, and energy equipment durability, existing desulphurisation strategies were qualitatively and quantitative reviewed. Preliminary work based on data obtained from the literatures suggested that the in-situ chemical dosing approach constituted the cheapest biogas desulphurisation technique, even after underlying uncertainties in the estimates were incorporated. Concerns are however raised with respect to system control costs and environmental costs, more so as continued supply of chemicals may be required for sustained desulphurisation and secondary pollution generation concerns, respectively. Finally, recognising that each desulphurisation approach considered may present different advantages and disadvantages, the integration of several technologies was proposed as an approach to counter the individual weaknesses of desulphurisation strategies, while also combining their strengths.

Author Contributions: Conceptualisation, Z.S. and O.V.O.; Methodology, O.V.O. and Z.S.; Writing—Original Draft Preparation, O.V.O. and Z.S.; Writing—Review and Editing, O.V.O. and Z.S.

Funding: This research received no external funding.

Conflicts of Interest: The authors declare no conflict of interest.

References

1. Okoro, O.V.; Sun, Z.; Birch, J. Catalyst-Free Biodiesel Production Methods: A Comparative Technical and Environmental Evaluation. *Sustainability* **2018**, *10*, 127. [CrossRef]
2. Okoro, O.V.; Sun, Z.; Birch, J. Techno-Economic Assessment of a Scaled-Up Meat Waste Biorefinery System: A Simulation Study. *Materials* **2019**, *12*, 1030. [CrossRef] [PubMed]
3. Tyagi, V.K.; Lo, S.-L. Energy and Resource Recovery from Sludge: Full-Scale Experiences. In *Environmental Materials and Waste: Resource Recovery and Pollution Prevention*; Academic Press: New York, NY, USA, 2016; pp. 221–244.
4. Farida, H.; Chang, Y.L.; Hirotsugu, K.; Abdul, A.H.; Yoichi, A.; Takeshi, Y.; Hiroyuki, D. Treatment of Sewage Sludge Using Anaerobic Digestion in Malaysia: Current State and Challenges. *Front. Energy Res.* **2019**, *7*, 19. [CrossRef]
5. Okoro, O.V.; Sun, Z.; Birch, J. Prognostic assessment of the viability of hydrothermal liquefaction as a post-resource recovery step after enhanced biomethane generation using co-digestion technologies. *Appl. Sci.* **2018**, *8*, 2290. [CrossRef]
6. Dinel, H.; Mathur, S.P.; Brown, A.; Lévesque, M. A Field Study of the Effect of Depth on Methane Production in Peatland Waters: Equipment and Preliminary Results. *J. Ecol.* **1988**, *76*, 1083–1091. [CrossRef]
7. Stams, A.J.M.; Elferink, S.J.W.H.O.; Westermann, P. Metabolic interactions between methanogenic consortia and anaerobic respiring bacteria. *Adv. Biochem. Eng. Biotechnol.* **2003**, *81*, 31–56. [PubMed]
8. Dai, Q.J.; Yi, L.I.; Fang, X.J. Review of desulfurization process for biogas. *IOP Conf. Ser. Earth Environ. Sci.* **2017**, *100*, 1–7.
9. Mackie, R.I.; Stroot, P.G.; Varel, V.H. Biochemical identification and biological origin of key odor components in livestock waste. *J. Anim. Sci.* **1998**, *76*, 1331–1342. [CrossRef]
10. Scheerer, U.; Haensch, R.; Mendel, R.R.; Kopriva, S.; Rennenberg, H.; Herschbach, C. Sulphur flux through the sulphate assimilation pathway is differently controlled by adenosine 5′-phosphosulphate reductase under stress and in transgenic poplar plants overexpressing γ-ECS, SO, or APR. *J. Exp. Biol.* **2009**, *61*, 609–622. [CrossRef]
11. Haghighatafshar, S. *Management of H$_2$S in Ananerobic Digestion of Enzyme Pretreated Macro Algae*; Lund University: Lund, Sweden, 2012.
12. Fardeau, M.L.; Ollivier, B.; Patel, B.K.C.; Dwivedi, P.; Ragot, M.; Garcia, J.L. Isolation and characterization of a thermophilic sulfate-reducing bacterium, *Desulfotomaculum thermosapovorans* sp. nov. *Int. J. Syst. Evol. Microbiol.* **1995**, *45*, 218–221. [CrossRef]
13. Goorissen, H.P.; Boschker, H.T.S.; Stams, A.J.M.; Hansen, T.A. Isolation of thermophilic *Desulfotomaculum* strains with methanol and sulfite from solfataric mud pools, and characterization of *Desulfotomaculum solfataricum* sp. nov. *Int. J. Syst. Evol. Microbiol* **2003**, *53*, 1223–1229. [CrossRef] [PubMed]
14. Stefanie, J.O.E.; Visser, A.; Pol, L.W.H.; Stams, A.J.M. Sulfate Reduction in Methanogenic Bioreactors. *FEMS Microbiol. Rev.* **1994**, *15*, 119–136.
15. Khoshnevisan, B.; Tsapekos, P.; Alfaro, N.; Díaz, I.; Fdz-Polanco, M.; Rafiee, S.; Angelidaki, I. A review on prospects and challenges of biological H$_2$S removal from biogas with focus on biotrickling filtration and microaerobic desulfurization. *Biofuel Res. J.* **2017**, *16*, 741–750. [CrossRef]
16. Sela-Adler, M.; Ronen, Z.; Herut, B.; Antler, G.; Vigderovich, H.; Eckert, W.; Sivan, O. Co-existence of Methanogenesis and Sulfate Reduction with Common Substrates in Sulfate-Rich Estuarine Sediments. *Front. Microbiol.* **2017**, *8*, 766. [CrossRef] [PubMed]

17. Enning, D.; Garrelfs, J. Corrosion of Iron by Sulfate-Reducing Bacteria: New Views of an Old Problem. *Appl. Environ. Microbiol.* **2014**, *80*, 1226–11236. [CrossRef] [PubMed]

18. Fauque, G.; Ollivier, B. Anaerobes: The sulphate-reducing bacteria as an example of metabolic diversity. In *Microbial Diversity and Bioprospecting*; ASm Press: Washington, DC, USA, 2004.

19. Hills, A.G. pH and the Henderson-Hasselbalch equation. *Am. J. Med.* **1973**, *55*, 131–133. [CrossRef]

20. Verma, N.K.; Khanna, S.K.; Kapila, B. *Comprehensive Chemistry*; Laxmi: New Delhi, India, 2010.

21. WHO. Hydrogen sulfide. In *Air Quality Guidelines for Europe Second Edition*; World Health Organisation: Copenhagen, Denmark, 2010; pp. 146–148.

22. Chaiprapat, S.; Charnnok, B.; Kantachote, D.; Sung, S. Bio-desulfurization of biogas using acidic biotrickling filter with dissolved oxygen in step feed recirculation. *Bioresour. Technol.* **2015**, *179*, 429–435. [CrossRef]

23. Horikawa, M.S.; Rossi, F.; Gimenes, M.L.; Costa, C.M.M.; Silva, M.G.C. chemical absorption of H_2S for biogas purification. *Braz. J. Chem. Eng.* **2004**, *21*, 415–422. [CrossRef]

24. Bai, Y.; Bai, Q. Subsea Corrosion and Scale. In *Subsea Engineering Handbook*, 2nd ed.; Gulf Professional Publishing: Oxford, UK, 2019; pp. 455–487.

25. CEN. *Standard in Development: BS EN 16723-2 Natural Gas and Biomethane for Use in Transport and Biomethane for Injection in the Natural Gas Network Part 2: Automotive Fuel Specifications*; European Commitee for Standardisation: Stockholm, Sweden, 2017.

26. Denyer, D.; Tranfield, D. *The Sage Handbook of Organizational Research Methods*; Sage: London, UK, 2009; pp. 671–689.

27. Müller-Langer, F.; Majer, S.; O'Keeffe, S. Benchmarking biofuels—A comparison of technical, economic and environmental indicators. *Energy Sustain. Soc.* **2014**, *4*, 1–14. [CrossRef]

28. Okoro, V.O. Scaled-Up Biodiesel Production from Meat Processing Dissolved Air Flotation Sludge: A Simulation Study. *AgriEngineering* **2019**, *1*, 17–43. [CrossRef]

29. Jenkins, S. Chemical Engineering Plant Cost Index: 2018 Annual Value. Available online: https://www.chemengonline.com/2019-cepci-updates-january-prelim-and-december-2018-final/ (accessed on 29 May 2019).

30. Turton, R.C.B.; Whiting, W.B.; Shaeiwitz, J.A. *Analysis Synthesis and Design of Chemical Processes*; Prentice Hall: Upper Saddle River, NJ, USA, 2009.

31. Michailos, S.; McCord, S.; Sick, V.; Stokes, G.; Styring, P. Dimethyl ether synthesis via captured CO_2 hydrogenation within the power to liquids concept: A techno-economic assessment. *Energy Convers. Manag.* **2019**, *184*, 262–276. [CrossRef]

32. Díaz, I.; Ramos, I.; Fdz-Polanco, M. Economic analysis of microaerobic removal of H_2S from biogas in full-scale sludge digesters. *Bioresour. Technol.* **2015**, *192*, 280–286. [CrossRef] [PubMed]

33. Deshusses, M.A.; Webster, T.S. Construction and Economics of a Pilot/Full-Scale Biological Trickling Filter Reactor for the Removal of Volatile Organic Compounds from Polluted Air. *J. Air Waste Manag. Assoc.* **2011**, *50*, 1947–1956. [CrossRef]

34. Nägele, H.J.; Steinbrenner, J.; Hermanns, G.; Holstein, V.; Haag, N.L.; Oechsner, H. Innovative additives for chemical desulphurisation in biogas processes: A comparative study on iron compound products. *Biochem. Eng. J.* **2017**, *121*, 181–187. [CrossRef]

35. Erdirencelebi, D.; Kucukhemek, M. Control of H_2S in full-scale anaerobic digesters using iron(III) chloride: Performance, origin and effects. *Water SA* **2018**, *44*, 176–183. [CrossRef]

36. Schäfer, F.; Dittrich-Zechendorf, M.; Leiker, M.; Pröter, J. Powdery ferric compounds for biogas process desulphurisation. *Landtechnik* **2017**, *72*, 39–48.

37. KRONOS. *Hydrogen Sulfide Elimination from Biogas*; KRONOS International Inc.: Dallas, TX, USA, 2014.

38. Jiang, H.; Li, T.; Stinner, W.; Nie, H. Selection of in-situ Desulfurizers for Chicken Manure Biogas and Prediction of Dosage. *Pol. J. Environ. Stud.* **2017**, *26*, 155–161. [CrossRef]

39. SevernWye. *Introduction to Production of Biomethane from Biogas: A Guide for England and Wales*; SevernWye Agency: Bangor, Wales, 2017.

40. Lupitskyy, R.; Alvarez-Fonseca, D.; Herde, Z.D.; Satyavolu, J. In-situ prevention of hydrogen sulfide formation during anaerobic digestion using zinc oxide nanowires. *J. Environ. Chem. Eng.* **2018**, *6*, 110–118. [CrossRef]

41. Allegue, L.B.; Hinge, J.H. *Biogas Upgrading Evaluation for H_2S Removal*; Danish Technology Institute: Taastrup, Denmark, 2014.

42. Krayzelova, L.; Bartacek, J.; Díaz, I.; Jeison, D.; Volcke, E.I.P.; Jenicek, P. Microaeration for hydrogen sulfide removal during anaerobic treatment: A review. *Rev. Environ. Sci. Biotechnol.* **2015**, *14*, 703–725. [CrossRef]

43. NYSERDA. *Assessment of Biochemical Process Controls for Reduction of Hydrogen Sulfide Concentrations in Biogas from Farm Digesters*; New York State Energy Research and Development Authority: New York, NY, USA, 2012.

44. Smith, J.A.; Carliell-Marquet, C.M. The digestibility of iron-dosed activated sludge. *Bioresour. Technol.* **2008**, *99*, 8585–8592. [CrossRef]

45. Al-Imarah, K.A.; Lafta, T.M.; Jabr, A.K.; Mohammad, A.N. Desulfurization for Biogas Generated by Lab Anaerobic Digestion unit. *J. Agric. Vet. Sci.* **2017**, *10*, 2319–2372.

46. Ofverstrom, S.; Dauknys, R.; Sapkaitė, I. The effect of iron salt on anaerobic digestion and phosphate release to sludge liquor. *Environ. Prot. Eng.* **2011**, *3*, 123–126.

47. Khanal, S.K.; Huang, J. Online Oxygen Control for Sulfide Oxidation in Anaerobic Treatment of High-Sulfate Wastewater. *Water Environ. Res.* **2006**, *78*, 397–408. [CrossRef] [PubMed]

48. Awe, O.W.; Zhao, Y.; Nzihou, A.; Minh, D.P.; Lyczko, N. A Review of Biogas Utilisation, Purification and Upgrading Technologies. *Waste Biomass Valoris.* **2017**, *8*, 267–283. [CrossRef]

49. Lein, C.; Wang, M.; Lin, W. Study on the removal of H_2S from biogas biomaterial using water scrubbing. *Trans. Tech. Publ.* **2017**, *723*, 599–603.

50. ToolBox, E. Solubility of Gases in Water. 2008. Available online: https://www.engineeringtoolbox.com/gases-solubility-water-d_1148.html (accessed on 23 April 2019).

51. Nie, H.; Jiang, H.; Chong, D.; Wu, Q.; Xu, C.; Zhou, H. Comparison of Water Scrubbing and Propylene Carbonate Absorption for Biogas Upgrading Process. *Energy Fuels* **2013**, *27*, 3239–3245. [CrossRef]

52. Dyment, J.; Watanasiri, S. *Acid Gas Cleaning Using DEPG Physical Solvents: Validation with Experimental and Plant Data*; Aspentech: Bedford, UK, 2015.

53. Petersson, A.; Wellinger, A. *Biogas Upgrading Technologies–Developments and Innovations. Task 37-Energy from Biogas and Landfill Gas*; EA Bioenergy: Aadorf, Switzerland, 2009.

54. Deshmukh, G.M.; Shete, A.; Pawar, D.M. Oxidative Absorption of Hydrogen Sulfide using Iron-chelate Based Process: Chelate Degradation. *J. Anal. Bioanal. Tech.* **2012**, *3*, 138. [CrossRef]

55. Frare, L.; Vieira, M.; Silva, M.; Pereira, N.; Gimenes, M. Hydrogen Sulfide Removal from Biogas Using Fe/EDTA Solution: Gas/Liquid Contacting and Sulfur Formation. *Environ. Prog. Sustain. Energy* **2009**, *29*, 34–41. [CrossRef]

56. Neumann, D.W.; Lynn, S. Oxidative absorption of H_2S and O_2 by iron chelate solutions. *Am. Inst. Chem. Eng.* **1984**, *30*, 62–69. [CrossRef]

57. Maia, D.C.S.; Niklevicz, R.R.; Arioli, R.; Frare, L.M.; Arroyo, P.A.; Gimenes, M.L.; Pereira, N.C. Removal of H_2S and CO_2 from biogas in bench scale and the pilot scale using a regenerable Fe-EDTA solution. *Renew. Energy* **2017**, *109*, 188–194. [CrossRef]

58. Krischan, J.; Makaruk, A.; Harasek, M. Design and scale-up of an oxidative scrubbing process for the selective removal of hydrogen sulfide from biogas. *J. Hazard. Mater.* **2012**, *215*, 49–56. [CrossRef] [PubMed]

59. Coppola, G.; Papurello, D. Biogas Cleaning: Activated Carbon Regeneration for H_2S Removal. *Clean Technol.* **2018**, *1*, 40–57. [CrossRef]

60. Magomnang, A.S.M.; Villanueva, E.P. Removal of Hydrogen Sulfide from Biogas using Dry Desulfurization Systems. In Proceedings of the International Conference on Agricultural, Environmental and Biological Sciences, Phuket, Thailand, 24–25 April 2014.

61. UNIDO. Biogas to Biomethane. 2017. Available online: https://www.biogas-to-biomethane.com/Download/BTB.pdf (accessed on 30 April 2019).

62. Cherif, H. *Study and Modeling of Separation Methods H2S from Methane, Selection of a Method Favoring H2S Valorization*; PSL Research University: Paris, France, 2016.

63. Rocca, M.A. Surface Science and Nanostructuring/Adsorption at Surfaces. 2012. Available online: http://www.fisica.unige.it/~{}rocca/Didattica/Surface%20Science%20and%20Nanostructuring/6%20adsorption%20at%20surfaces.pdf (accessed on 30 April 2019).

64. Al Mamun, M.R.; Torii, S. Removal of Hydrogen Sulfide (H_2S) from Biogas Using Zero-Valent Iron. *J. Clean Energy Technol.* **2015**, *3*, 428–432. [CrossRef]

65. Sitthikhankaew, R.; Predapitakkun, S.; Kiattikomol, R.; Pumhiran, S.; Assabumrungrat, S.; Laosiripojana, N. Comparative Study of Hydrogen Sulfide Adsorption by using Alkaline Impregnated Activated Carbons for Hot Fuel Gas Purification. *Energy Procedia* **2011**, *9*, 15–24. [CrossRef]

66. Louhichi, S.; Ghorbel, A.; Takfaoui, A.; Chekir, H.; Trabelsi, N.; Khemakhem, S. Alkaline activated carbon as adsorbents of hydrogen sulfide gases from chimney of phosphoric units. *J. Mater. Environ. Sci.* **2018**, *9*, 2686–2691.

67. Kwaśny, J.; Balcerzak, W.; Rezka, P. Application of zeolites for the adsorptive biogas desulfurisation. *Tech. Trans. Environ. Eng.* **2015**, 39–45. [CrossRef]

68. Micoli, L.; Bagnasco, G.; Turco, M. H_2S removal from biogas for fuelling MCFCs: New adsorbing materials. *Int. J. Hydrog. Energy* **2014**, *39*, 1783–1787. [CrossRef]

69. Feldbauer, S.L. *Steam Treating; Enhancing the Surface Properties of Metal Components*; Aabbott Furnace Company: St. Marys, PA, USA, 2003.

70. Wang, D. *Breakthrough Behavior of H2S Removal with an Iron Oxide Based CG-4 Adsorbent in a Fixed-Bed Reactor*; University of Saskatchewan: Saskatoon, SK, Canada, 2008.

71. Raabe, T.; Erler, R.; Kureti, S.; Krause, H. Oxygen Removal during Biogas Upgrading using iron-based Adsorbents. In Proceedings of the International Gas Union Research Conference, Copenhagen, Denmark, 17–19 September 2014.

72. Pourzolfaghar, H.; Ismail, M.; Izhar, S.; MagharehEsfahan, Z. Review of H_2S Sorbents at Low-Temperature Desulfurization of Biogas. *Int. J. Chem. Environ. Eng.* **2004**, *5*, 22–28.

73. TVT. *Biogas to Biomethane Technology Review*; Vienna University of Technology: Vienna, Austria, 2012.

74. Huertas, J.I.; Giraldo, N.; Izquierdo, S. *Removal of H2S and CO2 from Biogas by Amine Absorption, Mass Transfer in Chemical Engineering Processes*; InTech: Rijeka, Croatia, 2011.

75. Paolini, V.; Petracchini, F.; Guerriero, E.; Bencini, A.; Drigo, S. Biogas cleaning and upgrading with natural zeolites from tuffs. *Environ. Technol.* **2016**, *37*, 1418–1427. [CrossRef] [PubMed]

76. Thanapong, D. *Micro-Aeration for Hydrogen Sulfide Removal from Biogas*; Iowa State University: Iowa Ames, IA, USA, 2009.

77. Tang, Y.; Shigematsu, T.; Morimura, S.I.; Kida, K. The effects of micro-aeration on the phylogenetic diversity of microorganisms in a thermophilic anaerobic municipal solid-waste digester. *Water Res.* **2004**, *38*, 2537–2550. [CrossRef] [PubMed]

78. Li, Y.; Qiao, W. Transformations and impacts of ammonia and H_2S in anaerobic digestion reactors. In *Ananerobic Biotechnology: Environmental Protection and Resource Recovery*; Imperial College Press: London, UK, 2015; pp. 109–133.

79. Huber, B.; Herzog, B.; Drewes, J.E.; Koch, K.; Müller, E. Characterization of sulfur oxidizing bacteria related to biogenic sulfuric acid corrosion in sludge digesters. *BMC Microbiol.* **2016**, *16*, 153. [CrossRef] [PubMed]

80. Kleinjan, W. Biologically Produced Sulphur Particles and Polysulphide Sulphide Ions. Effects on a Biotechnological Process for the Removal of Hydrogen suphide from Gas Streams. Ph.D. Thesis, Wageningen Universiteit, Wageningen, The Netherlands, 2005.

81. Díaz, I.; Pérez, S.I.; Ferrero, E.M.; Fernández–Polanco, F. Effect of oxygen dosing point and mixing on the microaerobic removal of H_2S in sludge digesters. *Bioresour. Technol.* **2011**, *102*, 3768–3775. [CrossRef] [PubMed]

82. Duangmanee, T. Micro-Aeration for Hydrogen Sulphide Removal from Biogas. Ph.D. Thesis, Iowa State University, Ames, IA, USA, 2009. [CrossRef]

83. Guerrero, L.; Montalvo, S.; Huiliñir, C.; Campos, J.L.; Barahona, A.; Borja, R. Advances in the biological removal of sulphides from aqueous phase in anaerobic processes: A review. *Environ. Rev.* **2015**, *24*, 84–100. [CrossRef]

84. Díaz, I.; Lopes, A.C.; Perez, S.I.; Fdz-Polanco, M. Determination of the optimal rate for the microaerobic treatment ofseveral H_2S concentrations in biogas from sludge digesters. *Water Sci. Technol.* **2011**, *64*, 233–238. [CrossRef] [PubMed]

85. Van den Ende, F.P.; van Gemerden, H. Sulfide oxidation under oxygen limitation by a Thiobacillus thioparus isolated from a marine microbial mat. *FEMS Microbiol. Ecol.* **1993**, *13*, 69–77.

86. Jeníček, P.; Horejš, J.; Pokorná-Krayzelová, L.; Bindzar, J.; Bartáček, J. Simple biogas desulfurization by microaeration—Full scale experience. *Anaerobe* **2017**, *46*, 41–45. [CrossRef]

87. Zitomer, D.H.; Shrout, J.D. High-sulfate, high chemical oxygen demand wastewater treatment using aerated methanogenic fluidized beds. *Water Environ. Res.* **2000**, *72*, 90–97. [CrossRef]

88. Jenicek, P.; Koubova, J.; Bindzar, J.; Zabranska, J. Advantages of anaerobic digestion of sludge in microaerobic conditions. *Water Sci. Technol.* **2010**, *62*, 427–434. [CrossRef]

89. Karhadkar, P.P.; Audic, J.; Faup, G.M.; Khanna, P. Sulfide and sulfate inhibition of methanogenesis. *Water Res.* **1987**, *21*, 1061–1066. [CrossRef]

90. Parker, M.L. Method for Removing Hydrogen Sulfide from Sour Gas and Converting It to Hydrogen and Sulfuric Acid. Ph.D. Thesis, Stanford University, Stanford, CA, USA, 1 June 2010.

91. Valdés, F.; Camiloti, P.R.; Rodriguez, R.P.; Delforno, T.P.; Carrillo-Reyes, J.; Zaiat, M.; Jeison, D. Sulfide-oxidizing bacteria establishment in an innovative microaerobic reactor with an internal silicone membrane for sulfur recovery from wastewater. *Biodegradation* **2016**, *27*, 119–130. [CrossRef]

92. Pokorna-Krayzelova, L.; Bartacek, J.; Theuri, S.N.; Gonzalez, C.A.S.; Prochazk, J.; Volcke, E.I.P.; Jenicek, P. Microaeration through a biomembrane for biogas desulfurization: Lab-scale and pilot-scale experiences. *Environ. Sci. Water Res. Technol.* **2018**, *4*, 1190–1200. [CrossRef]

93. Ramos, I.; Pérez, R.; Fdz-Polanco, M. Microaerobic desulphurisation unit: a new biological system for the removal of H_2S from biogas. *Bioresour. Technol.* **2013**, *142*, 633–640. [CrossRef] [PubMed]

94. Lópeza, L.R.; Dorado, A.D.; Mora, M.; Gamisans, X.; Lafuente, J.; Gabriel, D. Modeling an aerobic biotrickling filter for biogas desulfurization through a multi-step oxidation mechanism. *Chem. Eng. J.* **2016**, *294*, 447–457. [CrossRef]

95. Montebello, A.M.; Mora, M.; López, L.R.; Bezerra, T.; Gamisans, X.; Lafuente, J.; Baeza, M.; Gabriel, D. Aerobic desulfurization of biogas by acidic biotrickling filtration in a randomly packed reactor. *J. Hazard. Mater.* **2014**, *280*, 200–208. [CrossRef] [PubMed]

96. Barbusiński, K.; Kalemba, K. Use of biological methods for removal of H_2S from biogas in wastewater treatment plants—A review. *Arch. Civ. Eng. Environ.* **2016**, *9*, 103–112. [CrossRef]

97. Soreanu, G.; Béland, M.; Falletta, P.; Ventresca, B.; Seto, P. Evaluation of different packing media for anoxic H_2S control in biogas. *Environ. Technol.* **2009**, *30*, 1249–1259. [CrossRef] [PubMed]

98. Gabriel, D.; Cox, H.H.J.; Deshusses, M.A. Conversion of Full-Scale Wet Scrubbers to Biotrickling Filters for Control at Publicly Owned Treatment Works. *J. Environ. Eng.* **2004**, *130*, 1110–1117. [CrossRef]

99. Qiu, X.; Deshusses, M.A. Performance of a monolith biotrickling filter treating high concentrations of H_2S from mimic biogas and elemental sulfur plugging control using pigging. *Chemosphere* **2017**, *186*, 790–797. [CrossRef]

100. Tayar, S.P.; Guerrero, R.B.C.; Hidalgo, L.F.; Bevilaqua, D. Evaluation of Biogas Biodesulfurization Using Different Packing Materials. *Chemengineering* **2019**, *3*, 27. [CrossRef]

101. Devinny, J.S.; Deshusses, M.A.; Webster, T.S. *Biofiltration for Air Pollution Control*; CRC-Lewis Publishers: Boca Raton, FL, USA, 1999.

102. Liu, D.H.F. *Environmental Engineers' Handbook*; CRC Press: Boca Raton, FL, USA, 1999.

103. Webster, T.S.; Devinny, J.S. Biofiltration of Odors, Toxics and Volatile Organic Compounds from Publicly Owned Treatment Works. *Environ. Prog.* **1996**, *15*, 141–147. [CrossRef]

104. Converse, B.M.; Schroeder, E.D.; Iranpour, R.; Cox, H.H.J.; Deshusses, M.A. Odor and Volatile Organic Compound Removal from Wastewater Treatment Plant Headworks Ventilation Air Using a Biofilter. *Water Environ. Res.* **2003**, *75*, 444–454. [CrossRef] [PubMed]

105. Dumont, E. H_2S removal from biogas using bioreactors: A review. *Int. J. Energy Environ.* **2015**, *6*, 479–498.

106. Fernández, M.; Ramírez, M.; Gómez, J.M.; Cantero, D. Biogas biodesulfurization in an anoxic biotrickling filter packed with open-pore polyurethane foam. *J. Hazard. Mater.* **2014**, *264*, 529–535. [CrossRef]

107. Almenglo, F.; Bezerra, T.; Lafuente, J.; Gabriel, D.; Ramírez, M.; Cantero, D. Effect of gas-liquid flow pattern and microbial diversity analysis of a pilot-scale biotrickling filter for anoxic biogas desulfurization. *Chemosphere* **2016**, *157*, 215–223. [CrossRef] [PubMed]

108. Adams, G.M.; Hargreaves, R.; Witherspoon, J.; Ong, H.; Burrowes, P.; Easter, C.; Dickey, J.; James, F.; MacPherson, L.; Porter, R.; et al. *Identifying and Controlling Municipal Wastewater Odor Phase I: Literature Search and Review*; IWA publishing: Alexandria, Egypt, 2003.

109. Schiavon, M.; Ragazzi, M.; Torretta, V.; Rada, E.C. Comparison between conventional biofilters and biotrickling filters applied to waste bio-drying in terms of atmospheric dispersion and air quality. *Environ. Technol.* **2015**, *37*, 975–982. [CrossRef]

110. Pathak, N.; Mahajan, P.V. Ethylene Removal from Fresh Produce Storage: Current Methods and Emerging Technologies. *Ref. Modul. Food Sci.* **2017**. [CrossRef]

111. Oyarzún, P.; Arancibia, F.; Canales, C.; Aroca, G.E. Biofiltration of high concentration of H_2S using *Thiobacillus thioparus. Process Biochem.* **2003**, *39*, 165–170. [CrossRef]

112. SU, J.J.; Chen, Y.J.; Chang, Y.C. A study of pilot scale biogas bio-filter system for utilisation on pigs farms. *J. Agric. Sci.* **2014**, *152*, 217–224. [CrossRef]

113. Groenestijn, J.W.V. *Biotechniques for Air Pollution Control: Past, Present and Future Trend*; University of La Coruña Publisher: La Coruña, Spain, 2005.

114. Koe, L.C.C. Evaluation of a pilot-sclea bioscrubber for the removal of H_2S. *Water Environ. J.* **2000**, *14*, 432–435. [CrossRef]

115. Koutinas, M.; Peeva, L.G.; Livingston, A.G. An attempt to compare the performance of bioscrubbers and biotrickling filters for degradation of ethyl acetate in gas streams. *J. Chem. Technol. Biotechnol.* **2005**, *80*, 1252–1260. [CrossRef]

116. Van-Groenestijn, J.W. Bioscrubbers. In *Bioreactors for Waste Gas Treatment. Environmental Pollution*; Springer: Dordrecht, The Netherlands, 2001; pp. 133–162.

117. LeCloirec, P.; Humeau, P. Bioscrubbers. In *Air Pollution Prevention and Control: Bioreactors and Bioenergy*; John Wiley & Sons: West Sussex, UK, 2013; pp. 139–153.

118. Behera, B.C.; Mishra, R.R.; Dutta, S.K.; Thatoi, H.N. Sulphur oxidising bacteria in mangrove ecosystem: A review. *Afr. J. Biotechnol.* **2014**, *13*, 2897–2907.

119. Tang, K.; Baskaran, V.; Nemati, M. Bacteria of the sulphur cycle: An overview of microbiology, biokinetics and their role in petroleum and mining industries. *Biochem. Eng. J.* **2009**, *44*, 73–94. [CrossRef]

120. Simon, J.; Kroneck, P.M.H. Chapter Two—Microbial Sulfite Respiration. In *Advances in Microbial Physiology*; Academic Press: Oxford, UK, 2013; pp. 45–117.

121. Lampe, D.G.; Zhang, T.C. Evaluation of sulfur-based autotrophic denitrification. In Proceedings of the Great Plains/Rocky Mountain Hazardous Substance Research Center (HSRC)/Waste-management Education & Research Consortium (WERC) JointConference on the Environment, Albuquerque, NM, USA, 21–23 May 1996.

122. Hayakawa, A.; Hatakeyama, M.; Asano, R.; Ishikawa, Y.; Hidaka, S. Nitrate reduction coupled with pyrite oxidation in the surface sediments of a sulfide-rich ecosystem. *JGR Biogeosci.* **2013**, *118*, 639–649. [CrossRef]

123. Dolejs, P.; Paclík, L.; Maca, J.; Pokorna, D.; Zabranska, J.; Bartacek, J. Effect of S/N ratio on sulfide removal by autotrophic denitrification. *Appl. Microbiol. Biotechnol.* **2015**, *99*, 2383–2392. [CrossRef] [PubMed]

124. Soreanu, G.; Béland, M.; Falletta, P.; Ventresca, B.; Seto, P. Laboratory pilot scale study for H_2S removal from biogas in an anoxic biotrickling filter. *Water Sci.* **2008**, *57*, 201–207. [CrossRef] [PubMed]

125. Khanongnuch, R.; Di Capua, F.; Lakaniemi, A.; Rene, E.R.; Lensa, P.N.L. H_2S removal and microbial community composition in an anoxic biotrickling filter under autotrophic and mixotrophic conditions. *J. Hazard. Mater.* **2019**, *367*, 397–406. [CrossRef] [PubMed]

126. Muñoz, R.; Malhautier, L.; Fanlo, J.; Quijano, G. Biological technologies for the treatment of atmospheric pollutants. *Int. J. Environ. Anal. Chem.* **2015**, *95*, 950–967. [CrossRef]

127. López, L.R.; Dorado, A.D.; Mora, M.; Prades, L.I.; Gamisans, X.; Lafuente, J.; Gabriel, D. Modelling biotrickling filters to minimize elemental sulfur accumulation during biogas desulfurization under aerobic conditions. In Proceedings of the 7th European Meeting on Chemical Industry and Environment, Tarragona, Spain, 10–12 June 2015.

128. Mezzari, M.P.; Da Silva, M.L.B. *Sulfide Removal from Biogas by Sulphur Oxidising Bacteria*; Internacional Sobre Gerenciamento Deresíduos Agropecuários e Agroindustriais: SÃO Pedro, São Paulo, Brazil, 2013.

129. Syed, M.; Soreanu, G.; Falletta, P.; Béland, M. Removal of hydrogen sulfide from gas streams using biological processes—A review. *Can. Biosyst. Eng.* **2006**, *48*, 2.

130. Cork, D.; Mather, J.; Maka, A.; Srnak, A. Control of oxidative sulfur metabolism of Chlorobium limola forma thiosulfatophilum. *Appl. Environ. Microbiol.* **1985**, *49*, 269–272.

131. Pokorna, D.; Zabranska, J. Sulfur-oxidizing bacteria in environmental technology. *Biotechnol. Adv.* **2015**, *33*, 1246–1259. [CrossRef]

132. Massé, A.; Pringault, O.; Wit, R. Experimental Study of Interactions between Purple and Green Sulfur Bacteria in Sandy Sediments Exposed to Illumination Deprived of Near-Infrared Wavelengths. *Appl. Environ. Microbiol.* **2002**, *68*, 2972–2981. [CrossRef] [PubMed]

133. Gerardi, M.H.; Lytle, B. Purple and Green Sulfur Bacteria. In *The Biology and Troubleshooting of Facultative Lagoons*; John Wiley & Sons: New York, NY, USA, 2015; pp. 73–76.

134. Burns, A.S.; Padilla, C.C.; Pratte, Z.A.; Gilde, K.; Regensburger, M.; Hall, E.; Dove, A.D.M.; Stewart, F.J. Broad Phylogenetic Diversity Associated with Nitrogen Loss through Sulfur Oxidation in a Large Public Marine Aquarium. *Appl. Environ. Microbiol.* **2018**, *84*, e01250-18. [CrossRef] [PubMed]

135. Pfenning, N. Phototropic green and purple bacteria: A comparative, systematic survey. *Ann. Rev. Microbiol.* **1977**, *31*, 275–290. [CrossRef] [PubMed]

136. Nicholls, D.G.; Ferguson, S.J. Photosynthetic Generators of Protonmotive Force. In *Bioenergetics*, 4th ed.; Academic Press: Oxford, UK, 2013; pp. 159–196.

137. Müller, J.; Overmann, J. Close interspecies interactions between prokaryotes from sulfureous environments. *Front. Microbiol.* **2011**, *2*, 146. [CrossRef] [PubMed]

138. Robertson, L.A.; Kuenen, J.G. Denitrification by obligate and facultative autotrophs. In *Autotrophic Microbiology and One-Carbon Metabolism Vol. 1*; Kluwer Academic Publishers: Dordrecht, The Netherlands, 2012; pp. 93–117.

139. Rawat, R.; Rawat, S. Colorless sulfur oxidizing bacteria from diverse habitats. *Adv. Appl. Sci. Res.* **2015**, *6*, 230–235.

140. Canfield, D.E.; Kristensen, E.; Thamdrup, B. The Sulfur Cycle. In *Advances in Marine Biology*; Academic Press: Cambridge, MA, USA, 2005; pp. 313–381.

141. Barak, Y.; Tal, Y.; Rijn, J. Light-Mediated Nitrite Accumulation during Denitrification by *Pseudomonas* sp. Strain JR12. *Appl. Environ. Microbiol.* **1998**, *64*, 813–817.

142. Qambrani, N.A.; Oh, S. Effect of Dissolved Oxygen Tension and Agitation Rates on Sulfur-Utilizing Autotrophic Denitrification: Batch Tests. *Appl. Biochem. Biotechnol.* **2013**, *169*, 181–191. [CrossRef]

143. Kuenen, J.G.; Robertson, L.A.; Gemerden, H. Microbial Interactions among Aerobic and Anaerobic Sulfur-Oxidizing Bacteria. In *Advances in Microbial Ecology: Volume 8*; Springer: Boston, MA, USA, 1985; pp. 1–54.

144. Gemerden, H. Production of elemental sulfur by green and purple sulfur bacteria. *Arch. Microbiol.* **1986**, *146*, 52–56. [CrossRef]

145. Holkenbrink, C.; Barbas, S.O.; Mellerup, A.; Otaki, H.; Frigaard, N.U. Sulfur globule oxidation in green sulfur bacteria is dependent on the dissimilatory sulfite reductase system. *Microbiology* **2011**, *157*, 1229–1239. [CrossRef]

146. Fenchel, T.; King, G.M.; Blackburn, T.H. Aquatic Sediments. In *Bacterial Biogeochemistry*, 3rd ed.; Elsevier: London, UK, 2012; pp. 121–142.

147. Garrido, M.M. *Characterisation of S-Oxidising Biomass through Respirametric Techniques under Anoxic and Aerobic Conditions*; Univesitat Autonoma de Barcelona: Barcelona, Spain, 2014.

148. Pringault, O.; Kuhlt, M.; Wit, R.; Caumettel, P. Growth of green sulphur bacteria in experimental benthic oxygen, sulphide, pH and light gradients. *Microbiology* **1998**, *144*, 1051–1061. [CrossRef]

149. Bharathi, P.A.L. Sulfur Cycle. In *Encyclopedia of Ecology*; Academic Press: Oxford, UK, 2008; pp. 3424–3431.

150. Mahmood, Q.; Zheng, P.; Cai, J.; Wu, D.; Hu, B.; Li, J. Anoxic sulfide biooxidation using nitrite as electron acceptor. *J. Hazard. Mater.* **2007**, *147*, 249–256. [CrossRef] [PubMed]

151. Wahlund, T.M.; Tabita, F.R. The reductive tricarboxylic acid cycle of carbon dioxide assimilation: initial studies and purification of ATP-citrate lyase from the green sulfur bacterium *Chlorobium tepidum*. *J. Bacteriol.* **1997**, *179*, 4859–4867. [CrossRef] [PubMed]

152. Rothschild, L.J. The evolution of photosynthesis.again? *Phil. Trans. R. Soc. B* **2008**, *363*, 2787–2801. [CrossRef] [PubMed]

153. Kuenen, J.G. *Colourless sulphur* bacteria and their role in the sulphur cycle. *Plant Soil* **1975**, *43*, 49–76. [CrossRef]

154. Okafor, N. *Environmental Microbiology of Aquatic and Waste Systems*; Springer Science: Asheville, NC, USA, 2011.

155. Chemengonline. Economic Indicators. Available online: https://www.chemengonline.com/mediakit/wp-content/uploads/2016/01/EconomicIndicatorSampleSept17.pdf (accessed on 29 May 2019).

156. Shelford, T.; Gooch, C.; Choudhury, A.; Lansing, S. A Technical Reference Guide for Dairy-Derived Biogas Production, Treatment and Utilization. 2019. Available online: https://enst.umd.edu/sites/enst.umd.edu/files/_docs/FarmerbiogashandbookFinal.pdf (accessed on 20 July 2019).

157. Connolly, E.L.; Guerinot, M.L. Iron stress in plants. *Genome Biol.* **2002**, 3. [CrossRef]

158. Fdz-Polanco, M.; Diaz, I.; Perez, S.I.; Lopes, A.C.; Fdz-Polanco, F. H_2S removal in the anaerobic digestion of sludge by micro-aerobic processes: Pilot plant experience. *Water Sci. Technol.* **2009**, *60*, 3045–3050. [CrossRef] [PubMed]

159. Jenicek, P.; Celis, C.A.; Krayzelova, L.; Anferova, N.; Pokorna, D. Improving products of anaerobic sludge digestion by microaeration. *Water Sci. Technol.* **2014**, *69*, 803–809. [CrossRef] [PubMed]

160. Nghiem, L.D.; Manassa, P.D.M.; Fitzgerald, S.K. Oxidation reduction potential as a parameter to regulate micro-oxygen injection into anaerobic digester for reducing H_2S concentration in biogas. *Bioresour. Technol.* **2014**, *173*, 443–447. [CrossRef]

161. Cano, P.I.; Colón, J.; Ramírez, M.; Lafuente, J.; Gabriel, D.; Cantero, D. Life cycle assessment of different physical-chemical and biological technologies for biogas desulfurization in sewage treatment plants. *J. Clean. Prod.* **2018**, *181*, 663–674. [CrossRef]

162. Santos-Clotas, E.; Cabrera-Codony, A.; Castillo, A.; Martín, M.J.; Poch, M.; Monclus, H. Environmental Decision Support System for Biogas Upgrading to Feasible Fuel. *Energies* **2019**, *12*, 1546. [CrossRef]

Review

Microbial Ecology of Biofiltration Units Used for the Desulfurization of Biogas

Sylvie Le Borgne [1,*] and **Guillermo Baquerizo** [2,3]

1 Departamento de Procesos y Tecnología, Universidad Autónoma Metropolitana- Unidad Cuajimalpa, 05348 Ciudad de México, Mexico
2 Irstea, UR REVERSAAL, F-69626 Villeurbanne CEDEX, France
3 Departamento de Ingeniería Química, Alimentos y Ambiental, Universidad de las Américas Puebla, 72810 San Andrés Cholula, Puebla, Mexico
* Correspondence: sylvielb@correo.cua.uam.mx; Tel.: +52–555–146–500 (ext. 3877)

Received: 1 May 2019; Accepted: 28 June 2019; Published: 7 August 2019

Abstract: Bacterial communities' composition, activity and robustness determines the effectiveness of biofiltration units for the desulfurization of biogas. It is therefore important to get a better understanding of the bacterial communities that coexist in biofiltration units under different operational conditions for the removal of H_2S, the main reduced sulfur compound to eliminate in biogas. This review presents the main characteristics of sulfur-oxidizing chemotrophic bacteria that are the base of the biological transformation of H_2S to innocuous products in biofilters. A survey of the existing biofiltration technologies in relation to H_2S elimination is then presented followed by a review of the microbial ecology studies performed to date on biotrickling filter units for the treatment of H_2S in biogas under aerobic and anoxic conditions.

Keywords: biogas; desulfurization; hydrogen sulfide; sulfur-oxidizing bacteria; biofiltration; biotrickling filters; anoxic biofiltration; autotrophic denitrification; microbial ecology; molecular techniques

1. Introduction

Biogas is a promising renewable energy source that could contribute to regional economic growth due to its indigenous local-based production together with reduced greenhouse gas emissions [1]. Biogas can be used for heat and electricity generation, and after an upgrading process, as a natural gas substitute or as transportation fuel. Biogas is obtained from the decomposition of urban, industrial, animal or agricultural organic wastes under anaerobic conditions, a process called anaerobic digestion (AD) [2].

Biogas is a mixture typically composed of methane (CH_4) (50–75%) and carbon dioxide (CO_2) (25–50%) along with hydrogen sulfide (H_2S), ammonia (NH_3), aromatic, organochlorinated or organofluorated compounds and water vapor [3]. It contains H_2S at significant concentrations ranging from 0.005 to 2% (v/v) (50–20,000 ppmv) depending on the raw material used and the conditions of the AD process [4] (Table 1).

Biogas upgrading refers to the removal of CO_2, H_2S, H_2O and other trace contaminants such as siloxanes, halocarbons, O_2 and N_2. The type of upgrading process depends on the final use of biogas [5]. The removal of H_2S, the most significant reduced sulfur compound in biogas, is necessary for environmental, technical and health reasons. The combustion of non-desulfurized biogas leads to the emission of SOx that are precursors of acid rain and the presence of H_2S provokes the corrosion of combustion engines. Moreover, this gas emits a very unpleasant rotten-egg like odor which is detectable at very low concentrations (0.00047 ppmv) [6] and is highly toxic at concentrations of 50 ppmv and lethal at 300 ppmv [7]. H_2S inhibits cellular respiration after entering the bloodstream where it binds to and inhibits the cytochrome C oxidase in complex IV, the terminal enzymatic complex of the mitochondrial respiratory chain, leading to pulmonary paralysis, sudden collapse and death [7].

Table 1. H_2S content in biogas produced by anaerobic digestion (AD) of different wastes. Adapted from [2].

Biogas From	H_2S (ppm)
Wastewater AD plants	0–4000
Household waste	72–648
Agrifood industry	288
Agricultural waste	2160–7200
Landfill sites	0–100
Natural gas [$]	1.1–5.9

[$] Although it is not produced by AD, the H_2S content of natural gas is shown for comparison.

Several physical/chemical and biological technologies are available for H_2S removal (desulfurization). Physicochemical technologies including absorption, adsorption, chemical oxidation and membrane separation have been traditionally used for desulfurization. However, most of these technologies are characterized by the intensive use of energy and chemicals with the associated increase in operational costs and environmental impact due to the generation of emissions and hazardous by-products that must be treated and disposed of [5,8]. These characteristics have led to an intensification of the research on biological alternatives for biogas desulfurization.

Biological technologies (biotechnologies) operate at low temperatures and pressures and are based on the ability of certain microorganisms (i.e., sulfur-oxidizing bacteria, SOBs) to oxidize H_2S to innocuous products such as elemental sulfur (S^0) and sulfate (SO_4^{2-}) in the presence of O_2 or nitrate (NO_3^-) as the final electron acceptor (see Section 2). Biological desulfurization processes have become more popular due to their advantages compared to conventional technologies, including low energy requirements, the generation of harmless by-products and low investment and operation costs. The final products are non-hazardous: SO_4^{2-} can be directly discharged to receiving water bodies while S^0 can be separated and recovered to be used as a raw material for industrial and agricultural purposes [9].

The biological removal of H_2S in biogas has been conducted in gas-phase biological filter reactors (biofilters) and in algal-bacterial photobioreactors using the O_2 photosynthetically produced by microalgae or, in situ, in the headspace of AD vessels through the injection of micro-quantities of O_2 to stimulate the growth and activity of SOBs [5]. However, the main bioprocess employed has been biofiltration due to its high H_2S removal efficiencies, up to 99–100% depending on the concentration of H_2S at the inlet, and experience in full-scale implementation for waste gas treatment [10]. The objective of this review was to gather and discuss the current knowledge on microbial ecology in biofiltration units used for the removal of H_2S from biogas. Microbial ecology is the study of microorganisms in their natural environment and how microorganisms interact with each other and with the environment. The two main components of microbial ecology are biodiversity and microbial activity studies [11].

The novelty of this review article is to focus on microbial aspects of biogas desulfurization and depict the specific biofiltration technologies that could be used to treat high and variable loads of H_2S in biogas during extended periods of time, considering that the treated biogas is used in different applications after desulfurization. Such conditions are different from those found in the classical applications of biofiltration technologies for odor control in wastewater treatment plants (WWTPs) or other industrial processes.

The biological sulfur cycle is briefly presented, and the main characteristics of SOBs related to biofiltration are then reviewed with emphasis on their morphological and physiological diversity and metabolic versatility. The molecular techniques that have been used to characterize bacterial communities in biofilters are then briefly presented followed by a review on the current knowledge on microbial communities in biofiltration units used for biogas desulfurization under aerobic and anoxic conditions. The present article is the first presenting a review on the microbial ecology aspects of biogas desulfurization.

2. The Biological Sulfur Cycle and the Sulfur-Oxidizing Bacteria

Sulfur is the tenth most abundant element in the Earth's crust [12]. It is mainly found in the lithosphere, hydrosphere and atmosphere, in terrestrial and deep-sea hot springs, volcanic areas, mines, caves, seawater, in the form of sulfides (H_2S, HS^- and S^{2-}), sulfate minerals (gypsum, $CaSO_4$), sulfide minerals (pyrite, FeS_2), S^0 and SO_4^{2-} [12,13]. Human activities have impacted directly or indirectly by increasing the atmospheric emissions of sulfur in the form of H_2S (eutrophic marshes, sewage systems, several industries) and SOx (burning of fossil fuels) [13]. Sulfur is also important in the biosphere where it is incorporated into amino acids and proteins, hormones, lipids and vitamins. Biomass, which includes living and dead organic matter, constitutes a minor, but actively cycled, reservoir of sulfur [12].

Figure 1 shows a simplified version of the biological sulfur cycle and sulfur reservoirs, and highlights the importance of microorganisms, especially prokaryotes, in the cycling of inorganic sulfur compounds.

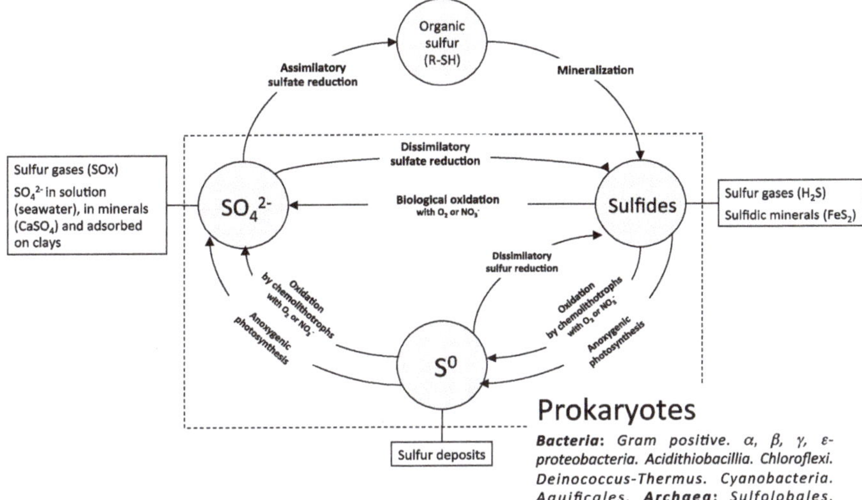

Figure 1. Sulfur reservoirs and the biological sulfur cycle. Adapted from [13].

SO_4^{2-} is the fully oxidized species of sulfur. It is reduced and assimilated by plants, fungi and bacteria to form amino acids and proteins (organic sulfur) that later become part of the sulfur-containing amino acids for humans and other animals. The decomposition of proteins during the mineralization of organic matter leads to the release of H_2S that reenters the cycle. H_2S is also produced by sulfate-reduction in anoxic habitats where sulfate-reducing bacteria (SRB) use SO_4^{2-} as the terminal electron acceptor for the oxidation of organic matter. This process is an anaerobic respiration, similar to aerobic respiration in which O_2 is used as the terminal electron acceptor, producing H_2O instead of H_2S as the metabolic by-product. SRB use SO_4^{2-} to generate energy in the cell, not to synthesize organosulfur compounds that become part of the cell material. In this sense, sulfate-reduction is a dissimilatory process. In the presence of SO_4^{2-} in biogas production processes, SRB can outcompete methanogenic bacteria for acetate and hydrogen (H_2), which are common substrates for both groups of bacteria, leading to the production of H_2S [14].

H_2S is a source of electrons for chemolithotrophic prokaryotes under aerobic and anaerobic conditions, and for phototrophic bacteria under strict anaerobic conditions when light is present. The phototrophic oxidation of sulfur is performed by green and purple sulfur bacteria. These bacteria incorporate carbon in the form of CO_2 or organic compounds using light energy, but instead of oxidizing H_2O to O_2 (oxygenic photosynthesis) they oxidize H_2S to S^0 and S^0 to SO_4^{2-} (anoxygenic photosynthesis). Dissimilative SOBs are chemolithotrophs that oxidize inorganic reduced sulfur compounds such as H_2S and S^0, using these compounds as electron donors for energy generation (ATP), typically with O_2 as the electron acceptor (aerobic respiration). Aerobic heterotrophic bacteria and fungi oxidize sulfur to thiosulfate or sulfate; however, the heterotrophic sulfur oxidation pathway has not been clearly elucidated yet [12] and even though fungi have been used in biofiltration applications, their application has mainly been for the elimination of volatile organic compounds (VOCs) [15]. More detailed information on sulfate-reduction and phototrophic or chemolithotrophic sulfur oxidation can be found elsewhere [11,16].

Figure 2 presents a brief summary of microbial trophic types and how cells convert carbon, energy and electrons to precursor metabolites, ATP, reducing power and new cell material. Dissimilatory sulfate-reduction, chemolithotrophic and phototrophic sulfur oxidation are forms of metabolism that only members of the domains *Bacteria* and *Archaea* can perform.

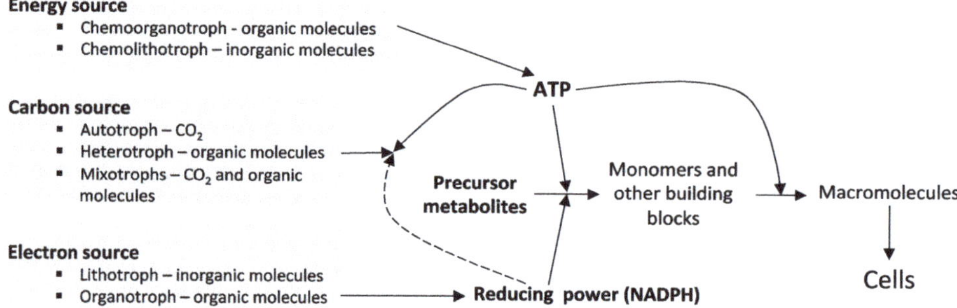

Figure 2. Microbial nutritional categories. Adapted from [17].

Chemolithotrophic SOBs are found in many natural and engineered environments where the sulfur cycle is active such as in marine sediments, sulfur springs, hydrothermal systems, sewage systems, anaerobic digesters and mines [18–21]. SOBs have been found in 2.5 billon year old fossils, prior to the Great Oxidation Event, when H_2S was an abundant energy source for microbial life [22]. H_2S is not toxic to these bacteria because their respiratory oxidase cytochrome *bd* is resistant to H_2S inhibition [23]. These bacteria are the base of the removal of H_2S from gases and airstreams by biofiltration and other environmental biotechnologies that have been extensively reviewed [10,24–26].

Currently, chemolithoautotrophic SOBs are the most predominant bacteria used for the biodegradation of H_2S [10,25,26] due to their versatility to operate in a wide range of environmental conditions (e.g., pH and temperature), low nutrients requirements, high sulfide tolerance, and slower growth rates than heterotrophs thus leading to less biomass accumulation. SOBs are morphologically, physiologically, phylogenetically and metabolically diverse.

For instance, some *Sulfurimonas* species (*Epsilonproteobacteria*) have a small cell size (0.66 × 2.1 µm) while *Thiomargarita* (*Gammaproteobacteria*), with an average cell diameter of 750 µm, is the largest bacterium discovered to date [27,28]. *Thiothrix* (*Gammaproteobacteria*) forms ensheathed filamentous multicellular structures that form rosettes under certain environmental conditions [29].

Thiobacillus sp. (*Betaproteobacteria*) and related genera (*Halothiobacillus*) are the best-studied chemolithotrophic SOBs and are known as the colorless sulfur bacteria in contrast to the phototrophic green and purple sulfur bacteria. The oxidation of H_2S and S^0 to SO_4^{2-} by *Thiobacillus* leads to the production of sulfuric acid (H_2SO_4) that acidifies the medium. Thus, many *Thiobacillus* species are acidophilic. The most acidophilic SOB, *Acidithiobacillus thiooxidans*, is typically found in acid mine drainage and corroded concrete (refer to Section 4). Haloalkaliphilic SOBs, *Thioalkalivibrio* and *Thioalkalimicrobium* sp. (*Gammaproteobacteria*) have been isolated from soda lakes and thrive at high pH (7.5 to 10.5) and high salt concentrations (1.5–4.3 M total Na^+) [30].

Many SOB species deposit the S^0 produced by the oxidation of H_2S in intracellular or extracellular granules for later use as an electron donor when H_2S is depleted. Bacterial sulfur globules show clear difference in the speciation of sulfur depending on the type of SOB, reflecting possible ecological and physiological properties [31]. *Thiomargarita namibiensis* forms intracellular sulfur rings and *Acidithiobacillus thiooxidans* extracellular polythionates.

Most SOBs are aerobic, however, some species of *Thiobacillus*, *Sulfurimonas* or *Thioalkalivibrio*, among others, can grow anaerobically with nitrate (NO_3^-) or nitrite (NO_2^-) as the electron acceptor. NO_3^- is sequentially reduced to NO_2^-, nitric oxide (NO), nitrous oxide (N_2O) and nitrogen gas (N_2), depending on the bacterial species and environmental conditions (refer to Section 3) [32]. This process is known as sulfur-oxidizing autotrophic denitrification. Most SOBs are obligate chemolithotrophs and can only use inorganic compounds as electron donors. Some SOBs are facultative chemolithotrophs and can grow either lithotrophically (autotrophs) or organotrophically. Bacteria that can simultaneously assimilate carbon CO_2 and organic sources are called mixotrophs (see Figure 2), for example *Thiothrix* sp. (refer to Section 4) [16].

No universal mechanism or pathway exists for sulfur oxidation in prokaryotes. Table 2 shows examples of the set of genes involved in sulfur oxidation in selected chemolithotrophic SOBs related to biofiltration applications. SOBs differ in the set of sulfur oxidation genes they contain, although some genes, but not all of them, are repeatedly found in different SOB species [33]. Well-studied examples include the complex SOx enzymatic system that catalyzes the oxidation of H_2S and S^0 to SO_4^{2-} while the DSR system is related to the formation of sulfur globules and the SQR enzyme to the oxidation of H_2S to S^0 [18]. Interestingly, some genes are found both in anaerobic and aerobic photo- and chemotrophic SOBs [18,33]. Some SOBs form symbiotic intra- or extracellular associations with marine invertebrates [34]. The deep-sea clams *Calyptogena* spp., which are found clustered near hydrothermal vents, harbor symbiotic chemolithoautotrophic SOBs in their gills' epithelial cells. These clams accumulate sulfide from the environment into their blood through their highly vascularized, muscular foot. The sulfide is transported via the blood to the gills where SOBs oxidize this reduced sulfur compound using it as an energy source for autotrophic growth and providing fixed carbon for the eukaryotic host. It has been shown that, in *Calyptogena*, key enzymes from five different sulfur oxidation pathways are equally expressed under three different environmental conditions (aerobic and semioxic) indicating that all pathways may function simultaneously to support intracellular endosymbiotic life [35]. This may be an advantage in an environment where the H_2S concentration rapidly fluctuates. No other reports are found concerning the expression of sulfur oxidation genes under different environmental conditions.

Table 2. Physiological characteristics and sulfur oxidation genes of some selected chemolithotrophic sulfur-oxidizing bacteria (SOBs). Adapted from [18].

SOB	Optimum pH Range	Anaerobic/Aerobic	Sulfur Oxidation Genes or Enzymes
Thiobacillus denitrificans	6.8–7.4	AN/AE	*sqr, fcc, sox* without *soxCD, dsr, apr*
Acidithiobacillus spp.	2–2.5	AN/AE *	*tet, tqo, sqr, sdo, tst, hdr, sox* without *soxCD* *
Thioalkalivibrio	9–10	AN/AE	*fcc, sox* without *soxCD, hdr, dsr* [&]
Sulfurimonas denitrificans	7	AN/AE	*sox, sqr* [#]

* *Acidithiobacillus thiooxidans* is only aerobic, taken from reference [36]. [&] Taken from references [37,38]. [#] Taken from reference [39].

3. Biofiltration Technologies

Since H_2S content in biogas may reach values up to 7000 ppmv, biological technologies should be able to withstand high and variable H_2S loads for extended periods of time. In the last few years, different biotechnologies addressing H_2S abatement in biogas and industrial waste gases have been proposed based on the activity of chemolithotrophic SOBs. In the case of large-scale application, most of them have been used for odor control in WWTPs and industrial processes. These waste gases are typically characterized by high flow rates and low contaminant concentrations that make them amenable to biological treatment [40].

Biological oxidation of H_2S using O_2 as the electron acceptor for the growth of colorless chemotrophic SOBs has been extensively reviewed [41]. *Thiobacillus* species together with *Sulfolobus*, *Thiovulum*, *Thiothrix* and *Thiospira* have been identified as representative genera [26]. The stoichiometry of H_2S oxidation reactions by chemolithotrophic SOBs in the presence of O_2 are as follows:

$$H_2S + 0.5\,O_2 \rightarrow S^0 + H_2O \qquad\qquad (-209.4\text{ kJ/reaction; } O_2/H_2S = 0.5) \qquad (1)$$

$$S^0 + 1.5\,O_2 + H_2O \rightarrow SO_4^{-2} + 2H^+ \qquad (-587\text{ kJ/reaction; } O_2/H_2S = 1.5) \qquad (2)$$

$$H_2S + 2O_2 \rightarrow SO_4^{-2} + 2H^+ \qquad\qquad (-798.2\text{ kJ/reaction; } O_2/H_2S = 2.0) \qquad (3)$$

S^0 is an intermediate compound formed under O_2-limited conditions, yielding less energy than the complete oxidation to SO_4^{2-}. The O_2/H_2S ratio will affect the final products obtained. Values slightly higher than the stoichiometry value, typically around 0.7, will lead to the formation of S^0 as the main final product, while ratios >1 will result in the significant formation of SO_4^{2-} [26].

Biological oxidation of H_2S can be also performed under anoxic conditions using oxidized forms of nitrogen (NO_3^- or NO_2^-) as the terminal electron acceptor instead of O_2 for the growth of autotrophic denitrifying SOBs. Some representative species are found in the following genera: *Thiobacillus, Thiomicrospira* and *Thiosphaera* [32,42–44]. Although most of the autotrophic denitrifying SOBs are facultative (e.g., the final electron acceptors can be O_2 or NO_3^-/NO_2^-), some strict anaerobes (e.g., *Sulfurimonas denitrificans*) have been reported [39,45]. In autotrophic denitrification coupled to sulfur oxidation, NO_3^- and/or NO_2^- are converted to N_2 through the following steps $NO_3^-\rightarrow NO_2^-\rightarrow NO\rightarrow N_2O\rightarrow N_2$ as described for traditional heterotrophic denitrification. *Thiobacillus denitrificans, Paracoccus versutus* and *Sulfurimonas denitrificans* can perform complete denitrification leading to the formation of N_2 while other species like *Thiobacillus thioparus* and *Thiobacillus delicatus* only reduce NO_3^- to NO_2^- [41,46]. Similarly to aerobic sulfur oxidation, the N/S ratio also affects the final products obtained [25,47]:

$$S^{2-} + 0.4\,NO_3^- + 2.4\,H^+ \rightarrow S^0 + 0.2\,N_2 + 1.2\,H_2O \qquad\qquad (191.0\text{ kJ/reaction; } N/S = 0.4) \quad (4)$$

$$S^{2-} + NO_3^- + 2\,H^+ \rightarrow S^0 + NO_2^- + H_2O \qquad\qquad (130.4\text{ kJ/reaction; } N/S = 1) \quad (5)$$

$$S^{2-} + 1.6\,NO_3^- + 1.6\,H^+ \rightarrow SO_4^{2-} + 0.8\,N_2 + 0.8\,H_2O \qquad (-743.9\text{ kJ/reaction; } N/S = 1.6) \quad (6)$$

$$S^{2-} + 4\,NO_3^- \rightarrow SO_4^{2-} + 4\,NO_2^- \qquad\qquad (-62.7\text{ kJ/reaction; } N/S = 4) \quad (7)$$

$$S^{2-} + 0.67\,NO_2^- + 2.67\,H^+ \rightarrow S^0 + 0.33\,N_2 + 1.33\,H_2O \qquad (-240.3\text{ kJ/reaction}) \quad (8)$$

$$S^{2-} + 2.67\,NO_2^- + 2.67\,H^+ \rightarrow SO_4^{2-} + 1.33\,N_2 + 1.33\,H_2O \qquad (-920.3\text{ kJ/reaction}) \quad (9)$$

As for aerobic sulfur oxidation, either S^0 or SO_4^{2-} will be formed depending on the N/S ratio. NO_3^- can be reduced to either NO_2^- or N_2. At a N/S ratio of 1.6, a complete oxidation of H_2S and reduction of NO_3^- to N_2 could be achieved with the highest energy yield [26]. However, the formation of S^0 and NO_2^- is frequently observed as the oxidation and reduction processes do not evolve to completion [10]. When N/S ≤ 0.4, the end products are S^0 and N_2 [47–49], If nitrates are stoichiometrically in excess (N/S > 4), the final products are nitrites, which accumulate and may also be used as electron acceptors [48,49].

Currently, the desulfurization of both biogas and industrial gas streams is mainly conducted in one of these bioreactors configurations: conventional biofilter, biotrickling filter (BTF) and suspended growth bioscrubber (Figure 3).

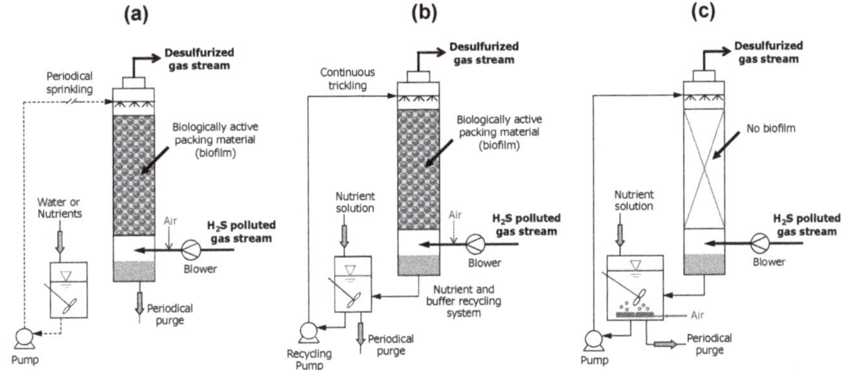

Figure 3. Biofiltration unit configurations: (**a**) Biofilter; (**b**) biotrickling filter (BTF); (**c**) Bioscrubber.

In biofilter and BTF systems, the contaminated gas stream is passed through a packed bed where the SOBs are immobilized in a biofilm. In conventional biofilters (Figure 3a), the packed bed is only periodically spread with water (or eventually nutrients) while in BTFs (Figure 3b) the packed bed is continuously trickled with a nutrient solution. The process is complex and involves several simultaneous physical, chemical and biological interactions. In conventional biofilters the pollutant (here H₂S) is transferred by absorption into the biofilm where diffusion and sulfur oxidation take place releasing S^0 or SO_4^{2-} (Figure 4). In BTFs, the difference is the presence of a continuous liquid phase that the pollutant first has to transfer into before transferring to the biofilm. Bioscrubbers are two-stage systems that consist of a gas scrubber and a biological reactor (Figure 3c). The H₂S is first transferred from the gas phase to an alkaline aqueous phase in an absorption column (gas scrubber) and then the resulting aqueous stream, containing the dissolved H₂S, is directed towards an agitated bioreactor where the H₂S is put into contact with SOBs (biological reactor). Thus, in bioscrubbers, the SOBs develop their activity in a stirred liquid while in biofilters and BTFs, the bioreaction takes place in a biofilm. The absorption and the biological oxidation reaction are physically separated in conventional bioscrubbers.

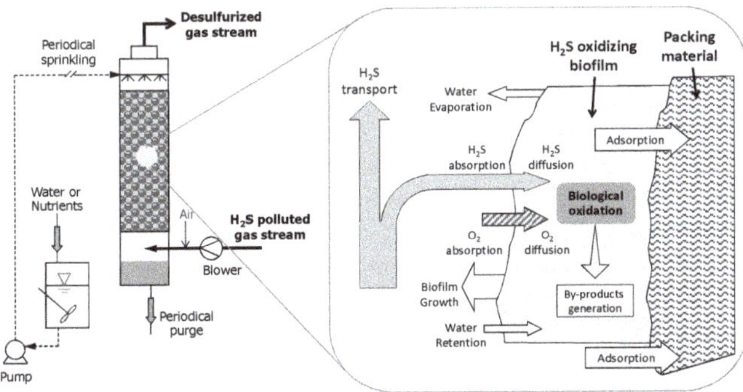

Figure 4. Schematic representation of the complex interactions that take place in biofilters.

Conventional biofilters have the simplest configuration. They have been used in large-scale gas applications for the control of H_2S and other odorous emissions from WWTPs and other industries [40]. The main benefits are the low operation cost, low-energy and chemical requirements, as well as high removal effciencies (REs), usually above 99% [50].

In conventional biofilters, the terminal electron acceptor is O_2 from the air since there is no continuously circulating liquid which could provide a constant supply of NO_3^- or NO_2^- as electron acceptors. Natural organic materials have normally been preferred for packing conventional biofilters (i.e., compost, peat, wood bark and soil among others) since they provide large specific areas with high porosity, low pressure loss, lightweight, low-cost, buffering and water-retaining capacity, intrinsic nutrient content as well as the presence of indigenous microbial consortia [51,52].

The drawbacks of conventional biofilters, especially under long-term operation, are (a) the accumulation of biomass and S^0 which may lead to bed clogging episodes (i.e., the reduction of inter-particle void space) causing preferential flow in the biofilter bed and pressure drop with the consequent reduction of the available mass transfer area; (b) acidification of the packing material due to the generation of SO_4^{2-} which leads to the formation of H_2SO_4, especially at the inlet area where the H_2S concentration and oxidation rate are higher, which may decrease the pH to values <1 causing inhibition of the microbial activity and decrease in the mass transfer rate into the biofilm; (c) compaction and degradation of the packing material provoking a reduction of media porosity and buffering capacity [53–57].

Conventional biofilters do not seem to be the most suitable technology for biogas desulfurization. The production of acid may result in the degradation of the packing material and the formation of small particles of degradation products contributing to bed clogging while, on the other hand, the accumulation of S^0 may also result in biofilter clogging [58], the main drawback of conventional biofilters, as mentioned above.

BTFs are a more sophisticated and controlled variation of conventional biofilters in which the aqueous phase is continuously trickled over the packed bed. Continuous trickling provides a better buffering capacity than in conventional biofilters and avoids excessive acidification in the packing bed through the continuous washout of SO_4^{2-} [59]. However, continuous nutrients supply and operation under higher loading rates than in conventional biofilters usually result in clogging caused by biomass growth and S^0 accumulation, both leading an increase in back pressure, bed channeling, formation of anaerobic zones and a decrease in RE [60,61]. Therefore, the control of biomass growth and S^0 accumulation arise as crucial operating parameters in BTFs. Different strategies have been used for limiting biomass overgrowth including the control of air supply to promote the oxidation of sulfide to SO_4^{2-}, the use of appropriate packing materials, the application of biomass predators, periodical bed backwashing and the control of nutrient supply [26,61]. Increasing the quantity of injected air results in higher O_2 levels favoring the formation of SO_4^{2-} thus alleviating the clogging problem due to S^0 accumulation [26]. However, this strategy reduces the off-gas quality and increases the risks of explosion. Optimizing the flow rate of the trickled liquid over the packing bed combined with recurrent draining and the application of a fresh trickling liquid have also been used to overcome clogging episodes during the treatment of VOCs in BTFs [61].

Another difference is the packing material, BTFs are usually packed with inert or synthetic materials including plastic rings, polyurethane foam (PUF), granular activated carbons, porous ceramics and lava rock and there are few applications reporting the use of natural materials [61–63]. Compaction and degradation of the packing material is therefore not a significant problem in BTFs. Moreover, synthetic packing beds maintain a relatively constant pressure drop, lower than natural materials. The open structure and high porosity of PUF results in a low pressure drop under conditions of high gas flow rates; additionally, PUF characteristics may also favor a faster biofilm formation in comparison to plastic materials and thus reducing the biofiltration start-up period [61].

Since the final product of the H_2S biological oxidation can be either S^0 or SO_4^{2-}, the O_2 mass transfer from gas to aqueous phase and biofilm (Figure 4) is one of the major parameters of this technology to ensure local stoichiometric level of O_2 in the biofilm [64].

Biogas desulfurization in BTFs through H_2S oxidation under chemolithoautotrophic denitrifying conditions has been recently recognized as a promising option [65–68]. The supply of NO_3^- or NO_2^- in the trickling liquid enables this biological process. Anoxic conditions eliminate the risk of explosion in CH_4/O_2 mixtures and dilution of the biogas with air. Moreover, there are no mass transfer limitations in the supply of NO_3^-/NO_2^-. In aerobic and anoxic BTFs, the generation of S^0 has to be controlled in order to avoid clogging effects, therefore the O_2/H_2S and N/S ratios are fundamental operational parameters.

Bioscrubbers have several advantages in comparison to biofilters and BTFs for biogas desulfurization. No O_2 is injected into the polluted gas stream, avoiding dilution effects and risks of explosion. Bioscrubbers can deal with fluctuating H_2S inlet loads, mostly because of the longer residence time of H_2S in the bioreactor [69] and easier management and control of the bioreactor operational condition, ensuring a stable and efficient operation [26]. Another advantage is that S^0 can be recovered from bioscrubbers and that there are no clogging problems. However, investment and operational costs are higher due to the presence of two separated operational units and large caustic consumption to maintain an efficient absorption, respectively. The Thiopaq® process (Paques, The Netherlands) and Sulfothane™ process (Biothane, USA) are bioscrubber-based systems that have been successfully developed at industrial large-scale for the aerobic desulfurization of biogas [7,25,26].

The main advantages and drawbacks of conventional biofilters, BTFs and bioscrubbers in relation to the treatment of H_2S in gas streams are summarized in Table 3. Among these three biotechnologies, the ability of BTFs to treat moderate to high H_2S loads has been demonstrated in numerous studies focused on the removal capacity and RE of BTFs treating H_2S at different loading rates, after shock loads, using different packing materials, under liquid flow patterns, as well as under different gas flow rates [70]. Hereafter, examples of studies combining the effect of different operational conditions and microbial diversity analyses, mainly in BTFs used in the removal of H_2S from biogas, are presented.

Table 3. Main advantages and drawbacks of biofiltration technologies used for H_2S abatement.

	Conventional Biofilters	BTFs	Bioscrubbers
Advantages	- Low investment and operational costs - Easy operation and maintenance - Effective removal at low H_2S concentrations	- Medium capital and low operating costs - Effective removal of H_2S at high concentrations - Easy process control (for example pH via trickled media) - Durability of the packing materials	- Treatment of high flow rates at different H_2S gas concentration - Operational stability - Low pressure drop
Drawbacks	- Bed clogging due to biomass growth and S^0 accumulation - Packing material compaction - Difficult to control pH drop due to SO_4^{2-} formation - Need of filter bed replacement	- Risk of bed clogging due to biomass growth and S^0 accumulation	- Relatively complex operation and maintenance - Secondary pollution generation (biomass and liquid waste streams)

4. Microbial Ecology Studies in Biofiltration Units for H_2S Removal

4.1. Molecular Techniques for Characterizing Bacterial Communities in Biofilters

Biofilters and BTFs make use of microorganisms embedded in a biofilm for the treatment of gaseous streams containing H_2S. Bacteria are the principal agents of sulfur oxidation. Since the performance of biofiltration processes depends on the activity and robustness of the bacterial communities involved, it is fundamental to characterize the bacterial populations that coexist in biofilters under different operational conditions in order to understand the sulfur cycling in these systems and propose potential improvements.

Classic microbiological methods based on cultivation favor the growth of certain groups of microorganisms. Artificial culture media, either liquid or solid, cannot reproduce the natural growth conditions present in biofilters, on the contrary, some nutrients may favor the growth of non-representative organisms only present in small amounts. Chemolithoautotrophic SOBs have slow growth rates and are usually closely associated in consortia containing several physiological and metabolic types of bacteria [71]. The limitations of culture techniques have been well documented, and it is now accepted that 99% of the microorganisms present in an environmental sample cannot be cultivated [72,73].

The advent of molecular techniques based on the small subunit of the ribosomal RNA gene (16S rRNA) has allowed the characterization of microbial communities from diverse environments such as humans, soils, oceans, engineered environments, bioreactors, bioremediation processes and, in the last 15 years, biofilters, among others. Phylogenetic analysis based on the 16S rRNA gene sequence is a gold standard for describing bacterial communities. This gene has been selected since the revolutionary studies of Carl Woese on the evolutionary history of cells (the three domains of life) and first applied by Norman Pace to the survey of natural microbial assemblages based on nucleic acid-based techniques [74,75]. This gene is ideal for bacterial phylogeny studies (evolution) and taxonomy purposes (classification) for the following reasons: (1) it is found in all prokaryotes, in single or multiple copies; (2) the function of this gene over time has not changed suggesting that changes in the sequence of this gene are a measure of evolution; (3) it offers sufficient resolution for discerning bacteria at the phylum to genus level; (4) it contains conserved regions for the design of general primers and probes; (5) it also has variable regions for the design of specific primers and probes and for phylogenetic studies; (6) it is large enough (1500 bp) for precise bioinformatic analyses [72–75]. Operational taxonomic units (OTUs) are used to categorize bacteria based on the similarity of their 16S rRNA gene sequences.

Millions of copies of these genes can be produced by using the polymerase chain reaction (PCR). The total DNA of a microbial community is used as the template in the PCR reaction. The mixture obtained after PCR amplification contains millions of copies of 16S rRNA sequences from every bacterium present in the original sample. These genes must then be separated, and their sequences determined in order to identify the corresponding bacteria. Several approaches have been used as fingerprinting methods, 16S rRNA clone libraries and direct amplicon sequencing without cloning using next generation sequencing technologies such as 454-pyrosequening, now discontinued, and Illumina platforms [76–78]. Among fingerprinting methods, denaturing gradient gel electrophoresis (DGGE) and single-strand conformation polymorphism (SSCP) allow the separation of 16S rRNA amplicons by denaturing and non-denaturing gel electrophoresis, respectively, while terminal restriction fragment length polymorphism (T-RFLP) allows the separation of fluorescently labelled restriction fragments of 16S rRNA amplicons. Fluorescence in situ hybridization (FISH) is a 16S rRNA targeted in situ hybridization technique used to label, visualize and enumerate whole cells in samples with PCR amplification [79].

4.2. Aerobic Biofiltration

The first study describing the dynamics of microbial communities in BTFs treating H_2S appears in 2005, in relation to odor abatement [80]. The bacterial communities present in a two-stage BTF system designed for the simultaneous removal of H_2S and dimethyl sulfide (DMS) are described using the DGGE fingerprinting method. The first BTF, aimed at removing H_2S, was inoculated with a pure culture of the acidophilic bacterium *Acidithiobacillus thiooxidans* and operated without pH control while the second BTF, aimed at removing DMS, was inoculated with the VS. strain of *Hyphomicrobium*, a neutrophilic bacterium, and operated at neutral a pH. This configuration was chosen to overcome the limitations encountered when treating gases containing mixtures of reduced sulfur compounds in which H_2S is preferentially degraded over organic sulfur compounds, especially at a low pH.

In the case of DMS (second BTF), the RE was found to be sensitive to lower values of the empty bed residence time (EBRT) and pH, H_2S overload and starvation times. However, the bacterial community in this BTF was found to be stable even on fluctuating DMS removal efficiencies. A high diversity was observed as shown by the multiple bands in the DGGE patterns and *Hyphomicrobium* VS. was no longer dominant, representing approximately only 10% of the established bacterial populations. A different trend was observed in the BTF inoculated with *Acidithiobacillus thiooxidans* (first BTF). The removal of H_2S was not affected by the operational conditions except when the H_2S load was increased from 1220 to 4037 ppmv. The pH, which was not controlled, maintained its value between 2 and 3 during the 117 days of operation and S^0 visibly accumulated. The bacterial diversity in this BTF was limited due to the low pH and only one prominent band corresponding to *Acidithiobacillus thiooxidans* was observed. This bacterium is the most acidophilic SOB and has been considered an ideal inoculum for the biofiltration of H_2S in biogas [50,81]. Its pH range for growth is between 0.5 and 5.5 with an optimum at pH 2–3 [36].

DGGE has also been used to study the evolution of bacterial populations in a two-stage BTF system for the simultaneous treatment of H_2S, methylmercaptan (MM), DMS and dimethyl disulfide (DMDS) for an odor abatement application [82]. The packing material consisted of polyurethane foam (PUF) cubes and the stream to be treated was generated in a compressor and enriched with H_2S and the above-mentioned organic reduced sulfur compounds. In the first BTF inoculated with *Acidithiobacillus thiooxidans* and operated at a low pH (2.5) for the removal of H_2S, the inoculated population remained constant at the different operation conditions tested (23 to 1320 ppmv of H_2S with and without different proportions of organic sulfur compounds). This result can also be explained by the low pH that restricted the colonization of this BTF by other bacteria. As reported by Sercu et al., (2005), the bacterial populations changed in the second BTF originally inoculated with *Thiobacillus thioparus* and operated under neutral conditions (pH 7.0), however the bacterial diversity in this BTF was low (one DGGE band observed) in spite of the permissive conditions of pH [80].

In the work of Sercu et al., (2005), the bacterial diversity in the recirculating liquid from the second BTF (for DMS removal) was also examined and it was observed that the planktonic bacterial community was quite different from the BTF community (biofilm), which highlights the importance of focusing on the biofilm [80].

A more extensive study on the diversity and spatial distribution of bacteria in a lab-scale BTF treating high H_2S loads (2000 ppmv) in a simulated biogas ("biogas mimic") was published in 2009 [83]. In addition to a fingerprinting molecular method (T-RFLP), a 16S rRNA gene clone library was constructed in order to further investigate the identity of the bacteria present in the BTF. Full length sequences of the 16S rRNA gene ($\approx$1500 pb) can be retrieved from clone libraries contrary to DGGE and T-RFLP that only analyze relatively small fragments from which phylogenetic affiliation of bacterial species cannot be precisely inferred. The BTF was packed with high density polypropylene grids (HD Q-Pack) and inoculated with a sulfur-oxidizing culture at pH 1.6 obtained from a full-scale biogas desulfurization column, enriched using Na_2S as the energy source and progressively acclimated to pH 6 [84]. An artificial biomass sample "representative" of the total community in the BTF was obtained by mixing samples (1:1) taken at the inlet and outlet parts of the reactor for constructing the gene clone library.

74% of the obtained sequences belonged to *Proteobacteria*, among which 49.4, 15.6 and 9.1% were *Gamma-*, *Beta-* and *Epsilonproteobacteria*, respectively. Of the 75 clones sequenced, 60% were related to SOB species, namely, *Thiothrix* sp. (*Gammaproteobacteria*); *Sulfurimonas denitrificans* (*Epsilonproteobacteria*); *Thiobacillus denitrificans*, *Thiobacillus sajanensis* and *Thiobacillus plumbophilus* (*Betaproteobacteria*). *Thiothrix* is a filamentous SOB that thrives in wastewaters characterized by high organic loads and elevated concentrations of low molecular weight fatty acids and reduced sulfur compounds, low dissolved oxygen (DO) concentrations and deficit in phosphorus and nitrogen [85]. The overgrowth of *Thiothrix* is related to activated sludge bulking in WWTPs. The presence of *Thiothrix* in the BTF was probably due to its ability to grow under heterotrophic, lithotrophic or mixotrophic conditions and use the organic matter, biomass and cellular debris accumulated in the BTF during its 6 months of operation. Sulfur-oxidizing mixotrophs can use reduced sulfur compounds as electron donors to produce ATP and CO_2 (autotrophy) or organic carbon (heterotrophy) to produce biomass. *Sulfurimonas* is a ubiquitous SOB found across the globe, in terrestrial and marine environments [86]. This bacterium is a strict chemoautotroph. *Sulfurimonas denitrificans* can use both sulfide and sulfur as electron donors and nitrate, nitrite and oxygen as electron acceptors [39]. *Thiobacillus* species are the typical colorless SOBs, with a strict chemolithoautotrophic metabolism. Among these, *Thiobacillus denitrificans* is a facultatively anaerobic bacterium, which can couple the oxidation of reduced sulfur compounds to denitrification [87].

Concerning the spatial distribution of bacterial populations, studied by T-RFLP, *Thiothrix* was more abundant in the outlet than in the inlet of the bioreactor. This finding is in agreement with niche differentiation patterns observed among sulfur-oxidizing populations in natural environments such as natural caves where three dimensions are critical, namely the sulfide, oxygen and water flow [88]. In such environments, *Thiothrix* is preferentially found in less turbulent or slowing flowing waters, with low sulfide and high oxygen. In BTFs, these conditions are met at the outlet. This is because, as in other plug flow types reactors, a gradient of H_2S and DO concentrations forms in BTFs, with higher H_2S/O_2 ratio in the inlet (bottom) and lower H_2S/O_2 ratio at the top (outlet). The conditions are therefore less severe in terms of H_2S concentration at the outlet.

Later on, the same group reported on the composition of the bacterial communities present in a similar laboratory scale BTF treating a biogas mimic, consisting of a mixture of H_2S and N_2, which was determined by using 16S rRNA clone libraries and FISH [89]. The H_2S concentration in the inlet was kept constant at 2000 ppm and the pH was controlled at 6.5–7 in the recirculating liquid. DO and pH were monitored online in the recirculating liquid. A 95% RE was achieved 3 days after inoculation of the BTF with an aerobic sludge from a municipal WWTP, indicating that this inoculum already contained a significant population of SOBs. An RE over 99% was obtained throughout the operation 10 days after start-up.

Equal amounts of biomass samples collected at days 42 and 189 from the inlet (bottom) and outlet (top) zones of the BTF were mixed to create "representative" samples from the total microbial community. The 16S rRNA clone libraries obtained 42 and 189 days after startup had a different composition indicating that the total bacterial communities present in the system changed throughout the operation, although the RE was above 99% throughout the study. Therefore, shuffling in the bacterial populations did not affect the performance of the BTF. A wide phylogenetic diversity was found in both libraries. After 42 and 189 days, 39 and 51% of the retrieved clones were affiliated with bacteria related to the sulfur cycle. The authors claim that using a biogas mimic is valid although it does not contain CH_4 or gaseous hydrocarbons since previous reports have indicated that the degradation of H_2S by lithotrophic bacteria is not affected by the presence of organic carbon sources. However, the presence of CH_4 in a real biogas may boost the growth of methanotrophic and heterotrophic bacteria that could compete for oxygen especially in areas of the bioreactor where DO is limiting. Diversity indexes indicated that bacterial diversity and evenness were lower at the longer operation time. So, a "simplification" or "metamorphosis" of bacterial communities towards more specific and dominant SOB communities was favored over long operation times in the presence of high H_2S loads.

Thiobacillus (*Betaproteobacteria*) was the most abundant genus detected (40%) at day 42 of operation while *Thiothrix* (*Gammaproteobacteria*) became the dominant genus (44%) after 189 days. Typical yellowish *Thiothrix*-like mats were observed at this time. The percentage of *Thiobacillus* sp. related bacteria decreased (7%) after 189 days of operation and the species detected at this time were different. As in the previous study, after some months of operation, more organic materials were present in the BTF and *Thiothrix* species displaced *Thiobacillus* sp. due to their ability to grow under heterotrophic, lithotrophic or mixotrophic conditions (see above for more information on *Thiothrix*). In contrast, *Thiobacillus* species are strictly chemolithotrophic. Other genera of SOBs were also detected, such as *Thiomonas*, *Sulfuricurvum*, *Acidithiobacillus* and *Halothiobacillus* at day 42 and *Sulfurimonas* at day 189.

Thiomonas species form a distinct phylogenetic cluster of SOBs capable of mixotrophic growth, which means they can incorporate organic substrates in the presence of an oxidable sulfur compound, this physiological feature clearly differentiates them from *Thiobacillus* species [90,91]. The potential of *Thiomonas* sp. for the removal of H_2S in a gas-bubbling reactor has been reported [92]. Some authors have reported that mixotrophic biofilters performed better than autotrophic biofilters based on H_2S RE after start-up, although most biofiltration processes have preferred autotrophic organisms due to their simplicity of operation and low biomass yield [92]. *Sulfuricurvum* is a strict chemolithotrophic SOB, that belongs to the *Epsilonproteobacteria*, a class that comprises other SOBs, such as *Sulfurimonas* (see above), with an important ecological and biogeochemical role in marine and terrestrial sulfidic habitats [93]. Low clone coverage was obtained (less than 49%) in both 16S rRNA libraries, which means that some SOB species were probably not detected.

Additionally, the authors used the FISH technique to follow the temporal and spatial dynamics of the SOBs in the BTF. For this, they used probes designed to detect the neutrophilic SOBs found in the clone libraries. Contrary to clone libraries, FISH is a quantitative technique for counting specific bacterial populations in environmental samples. It is not based on PCR, which is biased by the fact that rRNA operons have different copy numbers in different bacterial species. For example, according to the National Center for Biotechnology Information (NCBI) genome database [94], *Acidithiobacillus thiooxidans* and *Thiomonas intermedia* have one copy of the rRNA operon, *Thiothrix* sp., *Thiobacillus denitrificans* and *Halothiobacillus neapolitanus* have two copies and *Thiomonas intermedia* four copies. Additionally, in traditional PCR (end-point PCR), the proportion of dominant amplicons does not necessarily reflect the abundance of specific sequences, due to preferential amplification of certain sequences and accumulation of amplicons in the plateau phase of amplification, irrespective of their original abundance. In FISH with rRNA probes, single microbial cells are fluorescently stained and individually counted using a fluorescence microscope, independently of their rRNA content.

FISH revealed that *Thiothrix*, *Sulfurimonas denitrificans*, *Halothiobacillus neapolitanus*, *Thiobacillus denitrificans* and *Thiomonas* intermedia were present from day 0 at the inlet. *Thiomonas intermedia* notably increased at day 105 but was replaced by *Thiothrix* at day 189. A careful look at the monitored parameters at day 105 and 189 shows that the DO was globally lower at day 189, which may explain the difference in bacterial populations, although the DO was monitored in the liquid, not in the BTF, which may not be representative of local DO variations. The system was cleaned up after taking the sample at day 189 and, again at day 229, *Thiomonas intermedia* abundance increased, showing that this species may be a primary colonizer of the system in absence of significant quantities of organic matter. *Thiothrix* was more abundant at the outlet while the abundance of *T. intermedia* increased at the inlet compared to outlet, meaning that this species may be more adapted to high H_2S concentrations. After reactor clean up, *Thiomonas intermedia* was the most abundant SOB. As the pH was controlled in this system, the authors did not attribute the changes in microbial populations to pH but to DO, H_2S concentration and sulfur or organic matter accumulation. *Thiomonas intermedia* is a slightly acidophilic SOB that has been found in corroding sludge digesters and sewage systems [21,95].

In absence of pH control, the pH drops to very low values <2.0 in biofilters treating H_2S. The bacterial communities in pilot-scale BTFs packed with ceramic or volcanic rock, inoculated with an activated sludge from a municipal WWTP and operated at different EBRT (20, 15, 10 and 5 s) without pH control were studied [6]. The operation time at each EBRT was relatively short (15 days). The gas to be desulfurized proceeded from the WWTP with an H_2S load of 2.84 mg/m^3, corresponding to $\approx$2.0367 ppmv, a low load. The biomass samples for 16S rRNA gene pyrosequencing were collected in the bottom zone of the BTFs at the end of the first and third stages, corresponding to EBRTs of 20 and 10 s, respectively. The pH drops registered in the two BTFs were from 7.0 to 3.5 and 7.0 to 1.5 at EBRTs of 20 and 10 s, respectively. The volcanic rock BTF presented a 99% RE at any of the EBRTs tested while the RE of the ceramic BTF decreased from 94 to 60% at the lowest EBRT. The activity of the ceramic BTF was restored after washing with fresh medium, which means that it was more affected by pH than the volcanic rock BTF. Members of the *Thiomonas* genus were abundant in both BTFs and the abundance of this genus even increased under the most drastic conditions (a pH drop from 7 to 1.5) showing the acidophilic nature of this bacterium (see *Thiomonas* characteristics above). Only the volcanic BTF was colonized by *Acidithiobacillus thiooxidans* at the most extreme operating condition (EBRT of 10 s) with an abundance of 27.9%. In the ceramic BTF, the abundance of *Acidithiobacillus thiooxidans* was low (<2%) at any of the tested EBRTs. *Acidithiobacillus thiooxidans* was therefore clearly responsible for the performance of the volcanic rock BTF under acidic conditions. This bacterium has been proposed as ideal for inoculating biofilters operating under stable conditions of low pH as mentioned above. The influence of the nature of the packing material was evidenced here since the volcanic rocks, which have a higher porosity and specific surface area than the ceramic granules, allowed colonization by *Acidithiobacillus thiooxidans* and achieved high RE even at very low pH values. The ceramic BTF could not be colonized by *Acidithiobacillus thiooxidans*. It has been reported that the use of cell attachment promoters enhanced the performance of BTFs packed with PUF cubes and treating H_2S polluted airstreams [96]. The attachment of *Acidithiobacillus thiooxidans*, the dominant bacterium in these BTFs, to the PUF cubes was enhanced after coating the cubes with polyethyleneimine. It has been shown that *Acidithiobacillus thiooxidans* preferentially colonized concrete compared to glass in long incubation times, showing that biofilm formation by this bacterium may be material-dependent [97].

Concerning the influence of the packing material, there is only one report comparing the bacterial communities and the H_2S RE of two large-scale biofilters (not BTFs) packed with different inorganic materials, marble or pozzolan supplemented with limestone, and treating waste gases from the same WWTP [98]. The 16S rRNA gene clone libraries were constructed to compare the bacterial diversity in the two biofilters. In this study, the marble biofilter was very acidic (pH < 3), had a better H_2S removal performance and a higher bacterial diversity. The main OTUs detected in relation to sulfur oxidation were a betaproteobacterium from a sulfidic cave biofilm (not identified at the genus level) and *Thiobacillus sajanensis* with a reported optimum pH range of 6.8–9.5 [99]. Surprisingly, the biofilter packed with pozzolan, a material of volcanic origin, was less acidic (pH 5.7–6.8) and harbored different microbial populations such as the thermoacidophilic red algae *Cyanidium caldarium* and *Acidithiobacillus* sp. The authors attributed the increase of pH and lower H_2S RE to the presence of limestone.

The characterization of bacterial communities at three layers (bottom, middle and top) of a pilot scale peat biofilter treating gases emitted from a WWTP is described [40]. Although the system is not a BTF and the application is in odor abatement, the results are interesting in some aspects related to the influence of pH in biofilters at the pilot scale level. The molecular technique used here was SSCP. Peat is like a "self-inoculated" packing material with high organic content. The H_2S inlet concentrations vary from 227 to 1136 mg/m^3 (163–815 ppmv). During operation, the registered decrease of pH in the bottom, middle and upper zones was from 7.8 to 2.5, 7.5 to 3.3 and 7.3 to 6, respectively. These results are consistent with the highest pH drop at the inlet where H_2S biodegradation is the highest due to the accumulation of SO_4^{2-}. In this study, cell counts were performed at the different layers, however the media used were directed to heterotrophic bacteria and fungi not SOBs. The presence of fungi was noted at the bottom, consistent with the tolerance of fungi to acidic conditions. The bottom layer presented the highest bacterial diversity according to the obtained SSCP patterns and each layer was dominated by a few bacterial species, different in each layer, consistent with a stratification of the bacterial populations observed in other biofiltration systems, independently of the type of packing material and contaminant treated [100]. The few clones sequenced in this study allowed retrieving *Pseudomonas* and other heterotrophic bacteria instead of typical colorless SOBs. This is not surprising for this type of organic packing material.

The pH transition to acidic values drastically reduced the microbial diversity in a BTF packed with stainless steel Pall rings and inoculated with aerobic sludge from a local municipal WWTP [101]. A reference synthetic gas containing 2000 ppmv of H_2S was used. A gradual and controlled pH shift was established from pH 6.5 to 2.75 (in the recirculating liquid) between days 440 and 600 of operation. The total DNA was extracted from biomass collected at days 245 (neutral pH) and 586 (acidic pH) from three sampling ports (bottom 1/3, middle 2/3 and top 3/3) and mixed in equal proportions for 16S rRNA gene amplicons pyrosequencing. At an acidic pH, the community was enriched in *Gammaproteobacteria* where the acidic SOBs (*Acidithiobacillus* sp.) group instead of *Betaproteobacteria* where most neutrophilic SOBs are found. *Acidithiobacillus thiooxidans* was found in the communities developed under neutral conditions (4.1% of the OTU) together with *Thiomonas* and *Thiobacillus*. The abundance of *Acidithiobacillus* sp. increased up to 57.4% under acidic conditions. *Acidiphilium* was also detected (11.4%). This bacterium can grow under mixotrophic or chemolithotrophic conditions and is able to oxidize H_2S or elemental sulfur under oxygen limiting conditions. It is often found together with *Acidithiobacillus thiooxidans* in natural and engineered environments such as acid mine drainage and corroded concrete [102,103].

One of the most extensive studies, from a microbial ecology point of view, is perhaps the one described by Tu et al., (2015) [104]. In this study, two identical bench scale BTFs packed with volcanic rock, seeded with the same inoculum and acclimated for one month under steady state conditions at pH 4.0 (BTFa, a for acidic) or pH 7.0 (BTFn, n for neutral) were compared in terms of operational performance and microbial populations. The inoculum was obtained by mixing an aerobic sludge from a municipal WWTP and a microbial consortium enriched in the presence of thiosulfate from a sample collected in a landfill leachate treatment plant. The EBRTs tested were 60, 30 and 15 s for 14 h each, during which the H_2S load was gradually increased from 175 to 5858 mg/m^3, 169 to 5028 mg/m^3 and 69 to 1029 mg/m^3, respectively. Samples of the packing material were taken at the bottom (b), middle (m) and upper (u) layers of the BTFs at the end of each stage for pH measurements and MiSeq sequencing of 16s rRNA gene amplicons. The RE is also determined at the three layers for each BTF. The obtained results can be summarized in Table 4.

Table 4. Comparison of the operational behavior, pH and bacterial populations in two BTFs operated under acidic (a) and neutral conditions (n). EBRT: empty bed residence time.

Operational Behavior						
Biofilter	BTFa (pH 4)			BTFn (pH 7)		
EBRT (s)	60	30	15	60	30	15
Maximum RE at the inlet (%)	99	95	70	87.5	90	65
Average pH and Bacterial Populations						
Layer	Upper	Middle	Bottom	Upper	Middle	Bottom
pH	4.04	2.79	1.83	7.19	4.97	2.03
Abundance of *β-proteobacteria* (%)	20.0	32.9	23.6	25.2	29.9	19.1
Abundance of *β-proteobacteria* (%)	31.6	29.4	46.7	13.1	18.1	32.7

According to the RE values, BTFa had a better performance than BTFn. So, the most acidic condition (BTFa) generated the most effective BTF. Although the pH of the recirculating liquids was controlled at 4.0 and 7.0, gradients of pH formed into the two BTFs. As expected, in both BTFs, the pH was lower at the inlet layers where H_2S biodegradation is maximum due to the formation of SO_4^{2-}. The abundance of *Gammaproteobacteria*, where the acidic SOBs are found, was higher in the acidic BTF (46.7%). The predominant bacterial genus was *Acidithiobacillus thiooxidans*. The abundance of *Betaproteobacteria* (where neutrophilic SOBs are found) was between 19.1 and 32.9%. The most abundant genera were *Thiomonas*, *Thiobacillus* and *Halothiobacillus*. Considering these results, *Acidithiobacillus* was again found to be the key bacterium for H_2S biodegradation. The inoculation of biofilters with *Acidithiobacillus* species is proposed as an alternative to reduce acclimation times during H_2S biofiltration. Principal components analysis of diversity index values clearly separated the two BTFs and, for each BTF, the vertical stratification was clear with samples from the bottom layers being the most distant. This is another example of the influence of the operational conditions, pH and gradients of H_2S and O_2 that influence the composition and diversity of bacterial communities in BTFs for H_2S removal.

4.3. Anoxic Biofiltration

Fernandez et al. (2013) have reported the elimination of H_2S under anoxic conditions in a lab-scale BTF packed with Pall rings and operated at pH 7.4–7.5 [105]. The biogas was produced on-site in an upflow anaerobic sludge blanket (UASB) reactor and the H_2S content was gradually increased with a H_2S generation system. The inoculum consisted of biomass collected from a previous BTF packed with open polyurethane foam, operated under neutral pH conditions and initially seeded with nitrate-reducing bacteria and SOBs. Reduction of the methane and CO_2 content in the biogas were not observed during the 104 days of operation. The BTF was operated under nitrate limiting conditions (N/S ratio of 0.77 mol/mol), so the main product formed was elemental sulfur. The formation of this product could be reduced 25% if the N/S ratio was increased to 1.52 mol/mol. Analysis of the bacterial communities by DGGE was performed but it is not indicated at what time and how the sample was taken. Although the DDGE bands were not identified, a low bacterial diversity was observed. The same four bands were present at the inlet and at the outlet of the BTF while the recirculating liquid presented three bands, two of them also present in the BTF samples, meaning that some of the bacterial population were strongly attached to the packing material and that the nature of the inoculum, the anoxic conditions and the high H_2S loads (1400–14,000 ppmv) favored the establishment of a specialized bacterial community.

The effect of gas-liquid flow patterns on the performance and bacterial diversity and dynamics in a pilot-scale BTF treating a real biogas effluent from the AD of a WWTP was studied [65]. The BTF was packed with PUF cubes, inoculated with wastewater from the degritter-degreasing unit of the WWTP and operated for 415 days in five different operation modes: (1) day 1–297, counter-current flow with increasing H_2S loads; (2) day 298–360, co-current flow with increasing H_2S loads; (3) day 361–367, biogas supply cut; (4) steady state counter-current flow with liquid recirculation; and (5) steady state counter-current flow without liquid recirculation. Biomass samples were taken at the top and bottom of the BTF at day 343 (phase 2) and at the bottom at day 415 (phase 5) for massive pyrosequencing of short 16S rRNA gene amplicons. In counter-flow mode, biofilm growth rate is usually higher at the bottom part of BTFs where H_2S degradation is at its maximum. S^0 usually accumulates at the bottom where the ratio of electron acceptors (NO_3^- or O_2/H_2S) is the lowest. The slightly lower performance under co-current flow was attributed to the redistribution of biomass and S^0 along the packed bed as observed by visual inspection. This redistribution, however, promoted by alternate flow patterns, was favorable with respect to clogging and pressure drop problems. In addition, biogas supply cuts allowed the S^0 to be consumed by the microorganisms.

Concerning bacterial diversity temporal and spatial dynamics, a first observation is the high proportion of unclassified species found here, between 22 and 43% of the sequences, which is a drawback of the single-end sequencing of small 16S rRNA amplicons in the pyrosequencing technique. Presently, paired-end sequencing allows the sequencing of both ends of a fragment and generates high-quality, aligned sequence data, even with small fragments, in addition to high sequencing coverage. Bacterial diversity was similar at the bottom and top parts of the BTF operated under co-current flow (samples taken at day 343), with a strong dominance of *Proteobacteria*. *Sedimenticola* (*Gammaproteobacteria*) was the most dominant genus in both samples with a relative abundance of almost 49.5 and 44.2%. *Sedimenticola* is a versatile SOB comprising species that can grow lithoautotrophically under hypoxic and anaerobic conditions using a variety of anaerobic electron acceptors such as NO_3^-, (per)chlorate or chlorate and can also use organic compounds as a source of energy and electrons [106]. Bacteria from the *Rhizobiales* and *Rhodobacteraceae* were also detected at a significant abundance, 4.5 and 8.3%, respectively. *Rhizobiales* can fix N_2 while *Rhodobacteraceae* comprise chemoorgano- and photoheterotrophs putatively active in anoxic, nitrate-dependent sulfide oxidation [107]. The bacterial community structure drastically changed under counter-current flow without liquid recirculation. The community was more diverse, and most sequences were related to unclassified genera. It was therefore not possible to detect any SOBs. The performance of the BTF significantly decreased under single-pass flow (no liquid recirculation) at the two nitrate concentrations tested. The presence of so many unclassified sequences may be related to the growth of phototrophic eukaryotic microorganisms in this BTF since it was made of fiberglass, which shows some transparency to light. Moreover, the presence of nitrate may boost the growth of algae, which may further contribute to clogging problems. The presence of such organisms contaminates the samples with chloroplastic and mitochondrial 16S rRNA sequences which group along with unclassified sequences at the domain level. The degree of contamination depends on several factors, the DNA extraction and purification protocol, the primers and region of the 16S rRNA gene that was amplified and the bioinformatics pipeline used to analyze the data [108]. Overall, the obtained results indicate that the flow pattern has an influence on the composition of bacterial communities in BTFs and their vertical distribution, affecting the performance of the system and life span of the packed bed.

In the last year, several studies have reported the characterization of bacterial communities in anoxic BTFs for the removal of H_2S in gas streams. As expected and discussed above, the bacterial communities in anoxic BTFs are different from those of aerobic BTFs due to the use of different electron acceptors [109]. The packing material, open pore PUF (OPUF) or polypropylene Pall rings, did not influence the composition of the bacterial communities in BTFs seeded with the same inoculum (a sample from a previous OPUF biofilters) and treating 1400 to 14,600 ppmv H_2S concentrations as shown by the almost identical DGGE banding patterns [109]. The bacterial diversity was low probably because the inoculum had already been acclimated for biofiltration. The following genera and species were detected in relation to sulfur oxidation and autotrophic denitrification, *Thiobacillus thiophilus*, *Thiohalophilus* sp. and *Thiomonas intermedia*. *Thiobacillus thiophilus* has been described as an obligately chemolithoautotrophic and facultatively anaerobic bacterium, growing with either oxygen or nitrate as the electron acceptor [45]. *Thiohalophilus* sp. is an SOB that has been isolated under microoxic conditions and found to be capable of sulfur-driven anaerobic growth with NO_2^- [110]. As mentioned in Sections 2 and 3, NO_2^- is an intermediate in autotrophic and heterotrophic denitrification that can accumulate in biofilters when NO_3^- is used as the electron acceptor. The role of *Thiomonas intermedia* in sulfur oxidation has already been highlighted in aerobic biofilters, however, it is not clear if this bacterium can use NO_3^- as the electron acceptor. NO_2^- has been successfully used as the electron acceptor in a BTF treating a synthetic biogas containing H_2S concentrations of 952 to 3564 ppmv with a mineral medium as the recirculating liquid phase [111]. Although the bacteria were not identified, the authors report that the bacterial diversity was reduced during the progressive adaptation from NO_3^- to NO_2^-, however the DGGE banding patterns were similar, indicating that the same bacterial community was involved in sulfur-driven autotrophic denitrification with both electron acceptors. Finally, Khanongnuch et al. (2019) have just reported the anoxic desulfurization of a gas stream containing low H_2S concentrations (100–500 ppmv) at high EBRT values (3.5 min) using a synthetic nitrified wastewater as the recirculating liquid [68]. Using chemical sources of NO_3^- increases the operating costs of anoxic biofilters and the authors claim that using a nitrified wastewater as the trickling liquid would be a practical option if the H_2S treating BTF is located near a nitrification bioreactor. The obtained results indicated that H_2S elimination (RE >99%) via autotrophic denitrification was possible using nitrified wastewater and that *Thiobacillus* sp. was the only sulfur-oxidizing nitrate-reducing bacterial genus detected by DGGE. When the nitrified wastewater was amended with an organic compound to simulate the presence of residual organics, the H_2S RE drastically decreased to values between 60 and 80%. NO_3^- consumption increased due the growth of heterotrophic/mixotrophic denitrifying bacteria that probably outcompeted the autotrophic denitrifying SOBs leading to an increased accumulation of biomass. The detected heterotrophic denitrifiers were *Brevundimonas* and *Rhodocyclales*.

The main outcomes of microbial ecology studies in aerobic and anaerobic biofilters and BTFs for the abatement of H_2S in gas streams extensively reviewed in Sections 4.2 and 4.3, respectively, are summarized below in Table 5.

Table 5. Main outcomes of molecular microbial ecology studies on conventional biofilters and BTFs used for the removal of H_2S in odor abatement and biogas desulfurization applications. *A.t*: *Acidithiobacillus thiooxidans*; DGGE: denaturing gradient gel electrophoresis; DMS: dimethyl sulfide; FISH: fluorescence in situ hybridization; RSC: reduced sulfur compound; SSCP: single-strand conformation polymorphism; T-RFLP: terminal restriction fragment length polymorphism; UASB: upflow anaerobic sludge blanket; WWTP: wastewater treatment plant.

Application	Scale	Process Type	Inlet Gas	H_2S Load (ppmv)	Packing Material	Inoculum	Molecular Technique	Main Outcome from Microbial Ecology Studies	Ref.
Odor abatement	Lab	- Aerobic BTF - Operated at low pH - No pH control	Air supplemented with pure H_2S and DMS	1220–4037	Polyethylene carrier rings	*A.t*	DGGE	- Limited bacterial diversity - Predominance of *A.t*	[80]
Odor abatement	Lab	- Aerobic BTF - Operated at low pH - No pH control	Air supplemented with H_2S and other organic RSC	23–1320	Polyurethane foam	*A.t*	DGGE	- No change in bacterial diversity during operation - Predominance of *A.t*	[82]
Biogas desulfurization	Lab	- Aerobic BTF - Operation at pH 6	Biogas mimic (mixture of H_2S, N_2 and air)	2000	High density polypropylene grids	Culture from full-scale biogas desulfurization column (pH 1.6) adapted to pH 6	- T-RFLP - FISH - 16S gene clone library	- Presence of specialized bacterial populations - 60% of clones related to SOBs - Vertical stratification of bacterial populations along the length of the reactor	[83] [89]
Biogas desulfurization	Lab	Aerobic BTF with pH control at 6.5–7 in the recirculating liquid	Biogas mimic (mixture of H_2S, N_2 and air)	2000	High density polypropylene grids	Aerobic sludge from a municipal WWTP at pH		- Major abundance of facultative anaerobes at the inlet - Shifts in bacterial species during operation do not affect the RE	
Odor abatement	Pilot	- Aerobic BTF - No pH control	Waste gases from municipal WWTP	2.037	- Ceramic - Volcanic rocks	Activated sludge from municipal WWTP	- 16S rRNA gene amplicons pyrosequencing - 16S gene clone library	- Packing materials influence the composition of bacterial communities and the RE - Dominance of *A.t* at low pH	[6] [98]
Odor abatement	Large	- Conventional biofilter - No pH control	Used air from stabilizer or primary decanter of WWTP	>500	- Pozzolan added with calcareous material - Marble	Not described			
Odor abatement	Pilot	- Conventional biofilter - No pH control	Odorous gas from a WWTP	≅163–815	Peat	"Self-inoculated"	SSCP	- Vertical stratification of microbial populations related to pH gradient from top to bottom of the biofilter bed - Diversity of heterotrophic bacteria and presence of fungi instead of typical SOBs	[40]
Biogas desulfurization	Lab	- Aerobic BTF - Submitted to gradual pH shift from 6.5 to 2.75	Reference synthetic gas	2000	Steel pall rings	Aerobic sludge from a local municipal WWTP	16S rRNA gene amplicons pyrosequencing	- *A.t* dominates at low pH - The RE is not affected by the pH-dependent specialization of bacterial communities	[101]

Table 5. *Cont.*

Application	Scale	Process Type	Inlet Gas	H_2S Load (ppmv)	Packing Material	Inoculum	Molecular Technique	Main Outcome from Microbial Ecology Studies	Ref.
Odor abatement	Bench	- Aerobic BTFs - Operated at pH 4 or 7	Synthetic polluted gases generated by mixing H_2S vapors with fresh air	121–4200	Volcanic rock	Microbial consortium from biofilter treating landfill leachate waste gases + activated sludge from WWTP	MiSeq sequencing of 16S rRNA gene amplicons	- Vertical stratification of bacterial populations related to pH gradient in the BTFs - Bacterial community structure shaped by H_2S loading and pH of the recirculating nutrient solution	[104]
Biogas desulfurization	Lab	- Anoxic BTF at neutral pH - NO_3^- as electron acceptor	Biogas from UASB reactor	1400–14,000	Polypropylene Pall rings	Biomass from open-pore polyurethane foam of a previous BTF	DGGE	Specialized bacterial community	[105]
Biogas desulfurization	Pilot	- Anoxic BTF at neutral pH - NO_3^- as electron acceptor	Biogas split from anaerobic digester from WWTP	4490	Open-pore polyurethane foam	Wastewater from degritter-degreasing unit of WWTP	16S rRNA gene amplicons pyrosequencing	- The flow pattern shapes the composition of bacterial communities - No vertical stratification of bacterial communities observed under co-current flow - Less specialized bacterial communities with significant presence of heterotrophic, opportunistic species	[66]
Biogas desulfurization	Lab	- Anoxic BTFs at neutral pH - NO_3^- as electron acceptor	Biogas from UASB reactor	Not reported	- Open-pore polyurethane foam - Polypropylene Pall rings	Biomass from open-pore polyurethane foam of a previous BTF	DGGE	No influence of the packing material and operation time on bacterial diversity	[109]
Biogas desulfurization	Lab	- Anoxic BTFs at neutral pH - NO_3^- and NO_2^- as electron acceptors	Synthetic biogas (N_2 and H_2S)	710–3564	Polypropylene Pall rings	Not described	DGGE	Bacterial diversity reduced during the progressive adaptation from NO_3^- to NO_2^-	[111]
Odor abatement	Lab	- Anoxic BTF at neutral pH - Autotrophic or mixotrophic conditions - Synthetic nitrified wastewater as trickling liquid	Mixture of N_2 gas and H_2S generated using solutions of Na_2S and H_2SO_4	100–500	Polyurethane foam	Biofilm from a *Thiobacillus*-dominated lab-scale moving bed biofilm reactor	DGGE	Heterotrophic/mixotrophic denitrifying bacteria outcompete autotrophic denitrifying SOBs leading to an increased accumulation of biomass and decrease in the RE under mixotrophic conditions	[68]

5. Conclusions and Perspectives

Biofiltration appears to be a suitable biotechnology for the removal of H_2S from biogas. Different studies have shown that BTFs, contrary to conventional biofilters, are able to withstand high and variable loads of H_2S for extended periods of time, which is a prerequisite for application to the desulfurization of biogas. Large-scale applications of biofiltration for the removal of H_2S from gas streams, not necessarily biogas, make use of chemolithotrophic SOBs to oxidize H_2S to innocuous products such as SO_4^{2-} and S^0 in the presence of O_2. S^0 is formed as an intermediate. Although SO_4^{2-} is the preferred end product, S^0 generally accumulates in biofilters as the result of limited O_2 supply in the sulfur-oxidizing biofilm. S^0 accumulation causes clogging episodes, a main challenge for the application of BTFs, which can be alleviated by the periodical shutdown of BTF units to allow the biological oxidation of the accumulated S^0 to SO_4^{2-} in absence of H_2S.

Chemolithotrophic SOBs are part of the natural biogeochemical sulfur cycle where they play a fundamental role in the elimination of the H_2S produced by SRB in natural environments. Many of these bacteria are autotrophic, which is advantageous due to their low biomass production. They are very diverse from the morphological, phylogenetic, physiological and metabolic point of view, allowing them to thrive under a variety of environmental conditions. The ability of certain species of SOBs to use NO_3^- as the electron acceptor for the oxidation of H_2S is the base of anoxic biofiltration in BTFs. This new and less studied technology has been recognized as a promising option that would avoid the dilution of biogas and explosion risks due to the introduction of O_2.

As biofiltration is based on the activity of bacteria, it is important to reach a better understanding of the bacterial communities that populate BTFs under biogas desulfurization conditions. In the last 15 years, some studies describing the bacterial communities in aerobic and anoxic BTFs with the use of molecular biology tools based on the 16S rRNA gene sequence have been published in relation to biogas desulfurization (Table 5). These studies have shown that the environmental conditions imposed by the operational conditions have a direct impact on bacterial communities' diversity, structure and dynamics. Anoxic bacterial communities have been less studied and require a more intensive sequencing effort to more precisely determine the phylogenetic affiliation of the involved SOBs under different operational conditions. The robustness of the biological oxidation process is shown by the fact that the H_2S RE is maintained over extended periods of time in BTFs, even under fluctuating operational conditions as well as a change in the electron acceptor. Although a shift is observed in the bacterial communities' composition and structure, the performance of the BTF is maintained, showing the versatility, resilience and plasticity of bacterial sulfur-oxidizing communities. Vertical stratification of bacterial populations has been observed in aerobic BTFs, this spatial stratification is related to the H_2S/O_2 ratio along the packed bed. Extreme acidification due to the production of SO_4^{2-} that leads to the formation of H_2SO_4, does not inhibit the process as new SOB populations able to grow under extreme acidity progressively replace neutrophilic SOBs.

However, the 16S rRNA gene-based phylogenetic analysis does not identify the functional features of SOBs and there is still insufficient knowledge of the physiology and functional role of the key populations involved under different operational conditions. Recent advances in next generation sequencing technologies and bioinformatics has allowed the analysis of environmental metagenomes without PCR amplification to survey both the taxonomic and functional properties of microbial communities. The availability of complete genome sequences of different SOBs allows probes to be designed for sulfur oxidation genes for quantitative PCR applications. Finally, metatranscriptomic and metaproteomic approaches would allow a more complete picture of the metabolic role and activity of different SOBs to be obtained for better control and optimization of the biofiltration process.

Future research directions for biogas desulfurization should be focused on scaling-up the major outcomes found at the laboratory scale for anoxic BTFs to pilot-scale in order to determine the performance limits of these systems and the behavior of the involved microbial populations, especially for long-term operation using real biogas instead of biogas mimics. Additionally, aerobic and anoxic biotrickling filtration technologies should be submitted to detailed economic and environmental assessments using life cycle analysis-based approaches and, on this basis, compared to bioscrubber and physical/chemical technologies.

Author Contributions: S.L.B. proposed the structure and goal of the manuscript. S.L.B. and G.B. conceived and wrote the "1. Introduction", "2. The Biological Sulfur Cycle and the Sulfur-Oxidizing Bacteria" and "5. Conclusions and Perspectives" sections. G.B. conceived and wrote the "3. Biofiltration Technologies" section. S.L.B. conceived and wrote the "4. Microbial Ecology Studies in Biofiltration Units for H_2S Removal" section.

Funding: This research received no external funding.

Conflicts of Interest: The authors declare no conflict of interest.

References

1. Yentekakis, I.V.; Grammatiki, G. Biogas management: Advanced utilization for production of renewable energy and added-value chemicals. *Front. Environ. Sci.* **2017**, *5*, 7. [CrossRef]
2. Plugge, C.M. Biogas. *Microb. Biotechnol.* **2017**, *10*, 1128–1130. [CrossRef] [PubMed]
3. Biogas Renewable Energy. Available online: http://www.biogas-renewable-energy.info (accessed on 23 May 2019).
4. Arellano, L.; Dorado, A.D.; Fortuny, M.; Gabriel, D.; Gamisans, X.; González-Sánchez, A.; Hernández, S.; Lafuente, J.; Monroy, O.; Mora, M.; et al. *Purificación y Usos del Biogás*, 1st ed.; Universitat Autònoma de Barcelona: Barcelona, Spain, 2017; pp. 25–27.
5. Muñoz, R.; Meier, L.; Diaz, I.; Jeison, D. A review on the state-of-the-art of physical/chemical and biological technologies for biogas upgrading. *Rev. Environ. Sci. Biotechnol.* **2015**, *14*, 727–759. [CrossRef]
6. Li, J.; Ye, G.; Sun, D.; Sun, G.; Zeng, X.; Xu, J.; Liang, S. Performances of two biotrickling filters in treating H_2S-containing waste gases and analysis of corresponding bacterial communities by pyrosequencing. *Appl. Microbiol. Biotechnol.* **2012**, *95*, 1633–1641. [CrossRef] [PubMed]
7. Syed, M.; Soreanu, G.; Falletta, P.; Béland, M. Removal of hydrogen sulfide from gas streams using biological processes—A review. *Can. Biosyst. Eng.* **2006**, *48*, 2.
8. Niesner, J.; Jecha, D.; Stehlik, P. Biogas upgrading technologies: State of the art review in European region. *Chem. Eng. Trans.* **2013**, *35*, 517–522.
9. Chung, W.J.; Griebel, J.J.; Kim, E.T.; Yoon, H.; Simmonds, A.G.; Ji, H.J.; Dirlam, P.T.; Glass, R.S.; Wie, J.J.; Nguyen, N.A. The use of elemental sulfur as an alternative feedstock for polymeric materials. *Nat. Chem.* **2013**, *5*, 518–524. [CrossRef] [PubMed]
10. Vikrant, K.; Kailasa, S.K.; Tsang, D.C.W.; Lee, S.S.; Kumar, P.; Giri, B.S.; Singh, R.S.; Kim, K.-H. Biofiltration of hydrogen sulfide: Trends and challenges. *J. Clean. Prod.* **2018**, *187*, 131–147. [CrossRef]
11. Tortora, G.J.; Funke, B.R.; Case, C.L. *Microbiology: An Introduction*, 13th ed.; Pearson: New York, NY, USA, 2019.
12. Maier, R.M. Biogeochemical cycling. In *Environmental Microbiology*, 3rd ed.; Pepper, I.L., Gerba, C.P., Gentry, T.J., Eds.; Elsevier: San Diego, CA, USA, 2015; pp. 339–373.
13. Lens, P.N.; Kuenen, J.G. The biological sulfur cycle: Novel opportunities for environmental biotechnology. *Water Sci. Technol.* **2001**, *44*, 57–66. [CrossRef]
14. Moestedt, J.; Påledal, S.N.; Schnürer, A. The effect of substrate and operational parameters on the abundance of sulphate-reducing bacteria in industrial anaerobic biogas digesters. *Bioresour. Technol.* **2013**, *132*, 327–332. [CrossRef]
15. Vergara-Fernández, A.; Scott, F.; Moreno-Casas, P.; Revah, S. Removal of gaseous pollutants from air by fungi. In *Fungal Bioremediation: Fundamentals and Applications*; Tomasini, A., León-Santiesteban, H.H., Eds.; CRC Press, Taylor and Francis Group: Boca Raton, FL, USA, 2019; pp. 264–284.
16. Madigan, M.T.; Bender, K.S.; Buckley, D.H.; Stahl, D.A. *Brock Biology of Microorganisms*, 15th ed.; Pearson: New York, NY, USA, 2019.
17. Willey, J.M.; Sherwood, L.M.; Woolverton, C.J. *Prescott's Microbiology*, 10th ed.; Mc Graw Hill Education: New York, NY, USA, 2017.
18. Dahl, C.; Friedrich, C.; Kletzin, A. Sulfur oxidation in prokaryotes. In *Encyclopedia of Life Sciences (ELS)*; John Wiley & Sons, Ltd.: Chichester, UK, 2008; Available online: http://www.els.net/10.1002/9780470015902.a0021155 (accessed on 23 May 2019).

19. Tang, K.; Baskaran, V.; Nemati, M. Bacteria of the sulphur cycle: An overview of microbiology, biokinetics and their role in petroleum and mining industries. *Biochem. Eng. J.* **2009**, *44*, 73–94. [CrossRef]

20. Dopson, M.; Johnson, D.B. Biodiversity, metabolism and applications of acidophilic sulfur-metabolizing microorganisms. *Environ. Microbiol.* **2012**, *14*, 2620–2631. [CrossRef] [PubMed]

21. Huber, B.; Herzog, B.; Drewes, J.E.; Koch, K.; Müller, E. Characterization of sulfur oxidizing bacteria related to biogenic sulfuric acid corrosion in sludge digesters. *BMC Microbiol.* **2016**, *16*, 153. [CrossRef] [PubMed]

22. Czaja, A.-D.; Beukes, N.J.; Osterhout, J.T. Sulfur-oxidizing bacteria prior to the Great Oxidation Event from the 2.52 Ga Gamohaan Formation of South Africa. *Geology* **2016**, *44*, 983–986. [CrossRef]

23. Forte, E.; Giuffrè, A. How Bacteria Breathe in Hydrogen Sulfide-Rich Environments. *Biochem. Soc.* **2016**, *38*, 8–11. Available online: http://www.biochemist.org/bio/03805/0008/038050008.pdf (accessed on 23 May 2019).

24. Mohapatra, B.R.; Gould, W.D.; Dinardo, O.; Koren, D.W. An overview of the biochemical and molecular aspects of microbial oxidation of inorganic sulfur compounds. *CLEAN* **2008**, *36*, 823–829. [CrossRef]

25. Pokorna, D.; Zabranska, J. Sulfur-oxidizing bacteria in environmental technology. *Biotechnol. Adv.* **2015**, *33*, 1246–1259. [CrossRef]

26. Lin, S.; Mackey, H.R.; Hao, T.; Guo, G.; van Loosdrecht, M.C.M.; Chen, G. Biological sulfur oxidation in wastewater treatment: A review of emerging opportunities. *Water Res.* **2018**, *143*, 399–415. [CrossRef]

27. Labrenz, M.; Grote, J.; Mammitzsch, K.; Boschker, H.T.; Laue, M.; Jost, G.; Glaubitz, S.; Jürgens, K. *Sulfurimonas gotlandica* sp. nov., a chemoautotrophic and psychrotolerant epsilonproteobacterium isolated from a pelagic redoxcline, and an emended description of the genus *Sulfurimonas*. *Int. J. Syst. Evol. Microbiol.* **2013**, *63*, 4141–4148. [CrossRef]

28. Schulz, H.N.; Jorgensen, B.B. Big bacteria. *Annu. Rev. Microbiol.* **2001**, *55*, 105–137. [CrossRef]

29. Williams, T.M.; Unz, R.F.; Doman, J.T. Ultrastructure of *Thiothrix* spp. and "Type 021N" bacteria. *Appl. Environ. Microbiol.* **1987**, *53*, 1560–1570. [PubMed]

30. Sorokin, D.Y.; Kuenen, J.G. Haloalkaliphilic sulfur-oxidizing bacteria in soda lakes. *FEMS Microbiol. Rev.* **2005**, *29*, 685–702. [CrossRef] [PubMed]

31. Prange, A.; Chauvistré, R.; Modrow, H.; Hormes, J.; Trüper, H.G.; Dahl, C. Quantitative speciation of sulfur in bacterial sulfur globules: X-ray absorption spectroscopy reveals at least three different species of sulfur. *Microbiology* **2002**, *148*, 267–276. [CrossRef] [PubMed]

32. Shao, M.F.; Zhang, T.; Fang, H.H. Sulfur-driven autotrophic denitrification: Diversity, biochemistry, and engineering applications. *Appl. Microbiol. Biotechnol.* **2010**, *88*, 1027–1042. [CrossRef] [PubMed]

33. Ghosh, W.; Dam, B. Biochemistry and molecular biology of lithotrophic sulfur oxidation by taxonomically and ecologically diverse bacteria and archaea. *FEMS Microbiol. Rev.* **2009**, *33*, 999–1043. [CrossRef] [PubMed]

34. Stewart, F.J.; Dmytrenko, O.; Delong, E.F.; Cavanaugh, C.M. Metatranscriptomic analysis of sulfur oxidation genes in the endosymbiont of *Solemya velum*. *Front. Microbiol.* **2011**, *2*, 134. [CrossRef] [PubMed]

35. Harada, M.; Yoshida, T.; Kuwahara, H.; Shimamura, S.; Takaki, Y.; Kato, C.; Miwa, T.; Miyake, H.; Maruyama, T. Expression of genes for sulfur oxidation in the intracellular chemoautotrophic symbiont of the deep-sea bivalve *Calyptogena okutanii*. *Extremophiles* **2009**, *13*, 895–903. [CrossRef]

36. Wang, R.; Lin, J.Q.; Liu, X.M.; Pang, X.; Zhang, C.J.; Yang, C.L.; Gao, X.Y.; Lin, C.M.; Li, Y.Q.; Li, Y.; et al. Sulfur oxidation in the acidophilic autotrophic *Acidithiobacillus* spp. *Front. Microbiol.* **2019**, *9*, 3290. [CrossRef]

37. Muyzer, G.; Sorokin, D.Y.; Mavromatis, K.; Lapidus, A.; Clum, A.; Ivanova, N.; Pati, A.; d'Haeseleer, P.; Woyke, T.; Kyrpides, N.C. Complete genome sequence of "*Thioalkalivibrio sulfidophilus*" HL-EbGr7. *Stand. Genom. Sci.* **2011**, *4*, 23–35. [CrossRef]

38. Muyzer, G.; Sorokin, D.Y.; Mavromatis, K.; Lapidus, A.; Foster, B.; Sun, H.; Ivanova, N.; Pati, A.; D'haeseleer, P.; Woyke, T.; et al. Complete genome sequence of *Thioalkalivibrio* sp. K90mix. *Stand. Genom. Sci.* **2011**, *5*, 341–355. [CrossRef]

39. Sievert, S.M.; Scott, K.M.; Klotz, M.G.; Chain, P.S.; Hauser, L.J.; Hemp, J.; Hügler, M.; Land, M.; Lapidus, A.; Larimer, F.W.; et al. Genome of the Epsilonproteobacterial chemolithoautotroph *Sulfurimonas denitrificans*. *Appl. Environ. Microbiol.* **2008**, *74*, 1145–1156. [CrossRef] [PubMed]

40. Omri, I.; Bouallagui, H.; Aouidi, F.; Godon, J.J.; Hamdi, M. H₂S gas biological removal efficiency and bacterial community diversity in biofilter treating wastewater odor. *Bioresour. Technol.* **2011**, *102*, 10202–10209. [CrossRef]

41. Muyzer, G.; Kuenen, J.G.; Robertson, L.A. Colorless sulfur bacteria. In *The Prokaryotes*; Springer: New York, NY, USA, 2013; pp. 555–588.

42. Takai, K.; Suzuki, M.; Nakagawa, S.; Miyazaki, M.; Suzuki, Y.; Inagaki, F.; Horikoshi, K. *Sulfurimonas paralvinellae* sp. nov., a novel mesophilic, hydrogen- and sulfur-oxidizing chemolithoautotroph within the *Epsilonproteo-bacteria* isolated from a deep-sea hydrothermal vent polychaete nest, reclassification of *Thiomicrospira denitrificans* as *Sulfurimonas denitrificans* comb. nov. and emended description of the genus *Sulfurimonas*. *Int. J. Syst. Evol. Microbiol.* **2006**, *56*, 1725–1733. [PubMed]

43. Kellermann, C.; Griebler, C. *Thiobacillus thiophilus* sp. nov., a chemolithoautotrophic, thiosulfate-oxidizing bacterium isolated from contaminated aquifer sediments. *Int. J. Syst. Evol. Microbiol.* **2009**, *59*, 583–588. [CrossRef]

44. Shapleigh, J.P. Denitrifying prokaryotes. In *The Prokaryotes*; Springer: Berlin/Heidelberg, Germany, 2013; pp. 405–425.

45. Sorokin, D.Y.; Kuenen, J.G.; Jetten, M.S. Denitrification at extremely high pH values by the alkaliphilic, obligately chemolithoautotrophic, sulfur-oxidizing bacterium *Thioalkalivibrio denitrificans* strain ALJD. *Arch. Microbiol.* **2001**, *175*, 94–101. [CrossRef] [PubMed]

46. Robertson, L.A.; Kuenen, J.G. The colorless sulfur bacteria. In *The Prokaryotes. A Handbook on the Biology of Bacteria*; Springer: New York, NY, USA, 2006; pp. 985–1011.

47. Cardoso, R.B.; Sierra-Alvarez, R.; Rowlette, P.; Flores, E.R.; Gómez, J.; Field, J.A. Sulfide oxidation under chemolithoautotrophic denitrifying conditions. *Biotechnol. Bioeng.* **2006**, *95*, 1148–1157. [CrossRef]

48. Mahmood, Q.; Zheng, P.; Cai, J.; Wu, D.; Hu, B.; Islam, E.; Azim, M.R. Comparison of anoxic sulfide biooxidation using nitrate/nitrite as electron acceptor. *Environ. Prog. Sustain. Energy* **2007**, *26*, 169–177. [CrossRef]

49. An, S.; Tang, K.; Nemati, M. Simultaneous biodesulphurization and denitrification using an oil reservoir microbial culture: Effects of sulphide loading rate and sulphide to nitrate loading ratio. *Water Res.* **2010**, *44*, 1531–1541. [CrossRef] [PubMed]

50. Aita, B.C.; Mayer, F.D.; Muratt, D.T.; Brondani, M.; Pujol, S.B.; Denardi, L.B.; Hoffmann, R.; da Silveria, D.D. Biofiltration of H$_2$S-rich biogas using *Acidithiobacillus thiooxidans*. *Clean Technol. Environ. Policy* **2016**, *18*, 689–703. [CrossRef]

51. Aizpuru, A.; Khammar, N.; Malhautier, L.; Fanlo, J. Biofiltration for the treatment of complex mixtures of VOC–influence of the packing material. *Eng. Life Sci.* **2003**, *23*, 211–226. [CrossRef]

52. Khammar, N.; Malhautier, L.; Degrange, V.; Lensi, R.; Fanlo, J. Evaluation of dispersion methods for enumeration of microorganisms from peat and activated carbon biofilters treating volatile organic compounds. *Chemosphere* **2004**, *54*, 243–254. [CrossRef]

53. Mudliar, S.; Giri, B.; Padoley, K.; Satpute, D.; Dixit, R.; Bhatt, P.; Pandey, R.; Juwarkar, A.; Vaidya, A. Bioreactors for treatment of VOCs and odours—A review. *J. Environ. Manag.* **2010**, *91*, 1039–1054. [CrossRef] [PubMed]

54. Jaber, M.B.; Anet, B.; Amrane, A.; Couriol, C.; Lendormi, T.; Le Cloirec, P.; Cogny, G.; Fillières, R. Impact of nutrients supply and pH changes on the elimination of hydrogen sulfide, dimethyl disulfide and ethanethiol by biofiltration. *Chem. Eng. J.* **2014**, *258*, 420–426. [CrossRef]

55. Rabbani, K.A.; Charles, W.; Kayaalp, A.; Cord-Ruwisch, R.; Ho, G. Pilot-scale biofilter for the simultaneous removal of hydrogen sulphide and ammonia at a wastewater treatment plant. *Biochem. Eng. J.* **2016**, *107*, 1–10. [CrossRef]

56. Ben Jaber, M.; Couvert, A.; Amrane, A.; Le Cloirec, P.; Dumont, E. Removal of hydrogen sulfide in air using cellular concrete waste: Biotic and abiotic filtrations. *Chem. Eng. J.* **2017**, *319*, 268–278. [CrossRef]

57. Barbusinski, K.; Kalemba, K.; Kasperczyk, D.; Urbaniec, K.; Kozik, V. Biological methods for odor treatment—A review. *J. Clean. Prod.* **2017**, *152*, 223–241. [CrossRef]

58. Jin, J.; Veiga, M.C.; Kennes, C. Autotrophic deodorization o hydrogen sulfide in a biotrickling filter. *J. Chem. Technol. Biotechnol.* **2005**, *80*, 998–1004. [CrossRef]

59. Allegue, L.B.; Hinge, J. *Report: Biogas and Bio-Syngas Upgrading*; Danish Technological Institute: Taastrup, Denmark, 2012; pp. 1–97.

60. Iliuta, I.; Iliuta, M.C.; Larachi, F. Hydrodynamics modeling of bioclogging in waste gas treating trickle-bed bioreactors. *Ind. Eng. Chem. Res.* **2005**, *44*, 5044–5052. [CrossRef]

61. Rybarczyk, P.; Szulczynski, B.; Gebicki, J.; Hupka, J. Treatment of malodorous air in biotrickling filters: A review. *Biochem. Eng. J.* **2018**, *141*, 146–162. [CrossRef]

62. Cox, H.H.; Deshusses, M.A. Co-treatment of H_2S and toluene in a biotrickling filter. *Chem. Eng. J.* **2002**, *87*, 101–110. [CrossRef]

63. Lee, E.Y.; Lee, N.Y.; Cho, K.; Ryu, H.W. Removal of hydrogen sulfide by sulfate-resistant *Acidithiobacillus thiooxidans* AZ11. *J. Biosci. Bioeng.* **2006**, *101*, 309–314. [CrossRef]

64. Gabriel, D.; Deshusses, M.A.; Gamisans, X. Desulfurization of biogas in biotrickling filters. In *Air Pollution Prevention and Control*; John Wiley & Sons, Ltd.: Chichester, UK, 2013; pp. 513–523.

65. Almenglo, F.; Bezerra, T.; Lafuente, J.; Gabriel, D.; Ramirez, M.; Cantero, D. Effect of gas-liquid flow pattern and microbial diversity analysis of a pilot-scale biotrickling filter for anoxic biogas desulfurization. *Chemosphere* **2016**, *157*, 215–223. [CrossRef]

66. Lebrero, R.; Toledo-Cervantes, A.; Muñoz, R.; del Nery, V.; Foresti, E. Biogas upgrading from vinasse digesters: A comparison between an anoxic biotrickling filter and an algal-bacterial photobioreactor. *J. Chem. Technol. Biotechnol.* **2016**, *91*, 2488–2495. [CrossRef]

67. Ben Jaber, M.; Couvert, A.; Amrane, A.; Le Cloirec, P.; Dumont, E. Hydrogen sulfide removal from a biogas mimic by biofiltration under anoxic conditions. *J. Environ. Chem. Eng.* **2017**, *5*, 5617–5623. [CrossRef]

68. Khanongnuch, R.; Di Capua, F.; Lakaniemi, A.-M.; Rene, E.R.; Lens, P.N.L. H_2S removal and microbial community composition in an anoxic biotrickling filter under autotrophic and mixotrophic conditions. *J. Hazard. Mater.* **2019**, *367*, 397–406. [CrossRef]

69. Lebrero, R.; Bouchy, L.; Stuetz, R.; Muñoz, R. Odor assessment and management in wastewater treatment plants: A review. *Crit. Rev. Environ. Sci. Technol.* **2011**, *41*, 915–950. [CrossRef]

70. Gabriel, D.; Gamisans, X.; Muñoz, R. Technologies limiting gas and odor emissions. In *Innovative Wastewater Treatment and Resource Recovery Technologies: Impacts on Energy, Economy and Environment*; IWA Publishing: London, UK, 2017; pp. 233–254.

71. Okabe, S.; Ito, T.; Sugita, K.; Satoh, H. Succession of internal sulfur cycles and sulfur-oxidizing bacterial communities in microaerophilic wastewater biofilms. *Appl. Environ. Microbiol.* **2005**, *71*, 2520–2529. [CrossRef]

72. Amann, R.I.; Ludwig, W.; Schleifer, K.H. Phylogenetic identification and in situ detection of individual microbial cells without cultivation. *Microbiol. Rev.* **1995**, *59*, 143–169.

73. Pace, N.R. A molecular view of microbial diversity and the biosphere. *Science* **1997**, *276*, 734–740. [CrossRef]

74. Woese, C.R. Bacterial evolution. *Microbiol. Rev.* **1987**, *51*, 221–271.

75. Pace, N.R.; Stahl, D.A.; Olsen, G.J. Analyzing natural microbial populations by rRNA sequences. *ASM News* **1985**, *51*, 4–12.

76. Smalla, K.; Oros-Sichler, M.; Milling, A.; Heuer, H.; Baumgarte, S.; Becker, R.; Neuber, G.; Kropf, S.; Ulrich, A.; Tebbe, C.C. Bacterial diversity of soils assessed by DGGE, T-RFLP and SSCP fingerprints of PCR-amplified 16S rRNA gene fragments: Do the different methods provide similar results? *J. Microbiol. Methods* **2007**, *69*, 470–479. [CrossRef]

77. Reysenbach, A.L.; Wickham, G.S.; Pace, N.R. Phylogenetic analysis of the hyperthermophilic pink filament community in Octopus Spring, Yellowstone National Park. *Appl. Environ. Microbiol.* **1994**, *60*, 2113–2119.

78. Wen, C.; Wu, L.; Qin, Y.; Van Nostrand, J.D.; Ning, D.; Sun, B.; Xue, K.; Liu, F.; Deng, Y.; Liang, Y.; et al. Evaluation of the reproducibility of amplicon sequencing with Illumina MiSeq platform. *PLoS ONE* **2017**, *12*, e0176716. [CrossRef]

79. Yamaguchi, T.; Kubota, K. Visualization of microorganisms in bioprocesses. In *Optimization and Applicability of Bioprocesses*; Purohit, H.J., Kalia, V.C., Vaidya, A.N., Khardenavis, A.A., Eds.; Springer: Singapore, 2017; pp. 13–26.

80. Sercu, B.; Núñez, D.; Van Langenhove, H.; Aroca, G.; Verstraete, W. Operational and microbiological aspects of a bioaugmented two-stage biotrickling filter removing hydrogen sulfide and dimethyl sulfide. *Biotechnol. Bioeng.* **2005**, *90*, 259–269. [CrossRef]

81. Ramirez, M.; Gómez, J.M.; Cantero, D.; Páca, J.; Halecký, M.; Kozliak, E.I.; Sobotka, M. Hydrogen sulfide removal from air by *Acidithiobacillus thiooxidans* in a trickle bed reactor. *Folia Microbiol.* **2009**, *54*, 409–414. [CrossRef]

82. Ramírez, M.; Fernández, M.; Granada, C.; Le Borgne, S.; Gómez, J.M.; Cantero, D. Biofiltration of reduced sulphur compounds and community analysis of sulphur-oxidizing bacteria. *Bioresour. Technol.* **2011**, *102*, 4047–4053. [CrossRef]

83. Maestre, J.P.; Rovira, R.; Gamisans, X.; Kinney, K.A.; Kirisits, M.J.; Lafuente, J.; Gabriel, D. Characterization of the bacterial community in a biotrickling filter treating high loads of H_2S by molecular biology tools. *Water Sci. Technol.* **2009**, *59*, 1331–1337. [CrossRef]

84. Fortuny, M.; Baeza, J.A.; Gamisans, X.; Casas, C.; Lafuente, J.; Deshusses, M.A.; Gabriel, D. Biological sweetening of energy gases mimics in biotrickling filters. *Chemosphere* **2008**, *71*, 10–17. [CrossRef]

85. Henriet, O.; Meunier, C.; Henry, P.; Mahillon, J. Filamentous bulking caused by *Thiothrix* species is efficiently controlled in full-scale wastewater treatment plants by implementing a sludge densification strategy. *Sci. Rep.* **2017**, *7*, 1430. [CrossRef]

86. Han, Y.; Perner, M. The globally widespread genus Sulfurimonas: Versatile energy metabolisms and adaptations to redox clines. *Front. Microbiol.* **2015**, *6*, 989. [CrossRef]

87. Beller, H.R.; Chain, P.S.; Letain, T.E.; Chakicherla, A.; Larimer, F.W.; Richardson, P.M.; Coleman, M.A.; Wood, A.P.; Kelly, D.P. The genome sequence of the obligately chemolithoautotrophic, facultatively anaerobic bacterium *Thiobacillus denitrificans. J. Bacteriol.* **2006**, *188*, 1473–1488. [CrossRef]

88. Macalady, J.L.; Dattagupta, S.; Schaperdoth, I.; Jones, D.S.; Druschel, G.K.; Eastman, D. Niche differentiation among sulfur-oxidizing bacterial populations in cave waters. *ISME J.* **2008**, *2*, 590–601. [CrossRef]

89. Maestre, J.P.; Rovira, R.; Alvarez-Hornos, F.J.; Fortuny, M.; Lafuente, J.; Gamisans, X.; Gabriel, D. Bacterial community analysis of a gas-phase biotrickling filter for biogas mimics desulfurization through the rRNA approach. *Chemosphere* **2010**, *80*, 872–880. [CrossRef]

90. Moreira, D.; Amils, R. Phylogeny of *Thiobacillus cuprinus* and other mixotrophic Thiobacilli: Proposal for *Thiomonas* gen. nov. *Int. J. Syst. Evol. Microbiol.* **1997**, *47*, 522–528. [CrossRef]

91. Chen, X.G.; Geng, A.L.; Yan, R.; Gould, W.D.; Ng, Y.L.; Liang, D.T. Isolation and characterization of sulphur-oxidizing *Thiomonas* sp. and its potential application in biological deodorization. *Lett. Appl. Microbiol.* **2004**, *39*, 495–503. [CrossRef]

92. Chung, Y.-C.; Huang, C.-P.; Pan, J.R.; Tseng, C.-P. Comparison of autotrophic and mixotrophic biofilters for H_2S removal. *J. Environ. Eng. ASCE* **1998**, *124*, 362–367. [CrossRef]

93. Campbell, B.J.; Engel, A.S.; Porter, M.L.; Takai, K. The versatile *ε-proteobacteria*: Key players in sulphidic habitats. *Nat. Rev. Microbiol.* **2006**, *4*, 458–468. [CrossRef]

94. National Center for Biotechnology Information. Available online: https://www.ncbi.nlm.nih.gov (accessed on 23 May 2019).

95. Okabe, S.; Odagiri, M.; Ito, T.; Satoh, H. Succession of sulfur-oxidizing bacteria in the microbial community on corroding concrete in sewer systems. *Appl. Environ. Microbiol.* **2007**, *73*, 971–980. [CrossRef]

96. Goncalves, J.J.; Govind, R. Enhanced biofiltration using cell attachment promotors. *Environ. Sci. Technol.* **2009**, *43*, 1049–1054. [CrossRef]

97. Dec, W.; Cwalina, B.; Michalska, J.; Merkuda, D. Growth of *Acidithiobacillus thiooxidans* biofilm on glass, concrete and stoneware. *Solid State Phenom.* **2015**, *227*, 286–289. [CrossRef]

98. Chouari, R.; Dardouri, W.; Sallami, F.; Ben Rais, M.; Le Paslier, D.; Sghir, A. Microbial analysis and efficiency of biofiltration packing systems for hydrogen sulfide removal from wastewater off gas. *Environ. Eng. Sci.* **2015**, *32*, 121–128. [CrossRef]

99. Dul'tseva, N.M.; Turova, T.P.; Spiridonova, E.M.; Kolganova, T.V.; Osipov, G.A.; Gorlenko, V.M. *Thiobacillus sajanensis* sp. nov., a new obligately autotrophic sulfur-oxidizing bacterium isolated from Khoito-Gol hydrogen-sulfide springs, Buryatia. *Mikrobiologiia* **2006**, *75*, 670–681.

100. Lebrero, R.; Rodríguez, E.; Estrada, J.M.; García-Encina, P.A.; Muñoz, R. Odor abatement in biotrickling filters: Effect of the EBRT on methyl mercaptan and hydrophobic VOCs removal. *Bioresour. Technol.* **2012**, *109*, 38–45. [CrossRef]

101. Montebello, A.M.; Bezerra, T.; Rovira, R.; Rago, L.; Lafuente, J.; Gamisans, X.; Campoy, S.; Baeza, M.; Gabriel, D. Operational aspects, pH transition and microbial shifts of a H_2S desulfurizing biotrickling filter with random packing material. *Chemosphere* **2013**, *93*, 2675–2682. [CrossRef]

102. Norlund, K.L.; Southam, G.; Tyliszczak, T.; Hu, Y.; Karunakaran, C.; Obst, M.; Hitchcock, A.P.; Warren, L.A. Microbial architecture of environmental sulfur processes: A novel syntrophic sulfur-metabolizing consortia. *Environ. Sci. Technol.* **2009**, *43*, 8781–8796. [CrossRef]

103. Li, X.; Kappler, U.; Jiang, G.; Bond, P.L. The ecology of acidophilic microorganisms in the corroding concrete sewer environment. *Front. Microbiol.* **2017**, *8*, 683. [CrossRef]

104. Tu, X.; Li, J.; Feng, R.; Sun, G.; Guo, J. Comparison of removal behavior of two biotrickling filters under transient condition and effect of pH on the bacterial communities. *PLoS ONE* **2016**, *11*, e0155593. [CrossRef]

105. Fernández, M.; Ramírez, M.; Pérez, R.M.; Gómez, J.M.; Cantero, D. Hydrogen sulphide removal from biogas by an anoxic biotrickling filter packed with Pall rings. *Chem. Eng. J.* **2013**, *225*, 456–463. [CrossRef]

106. Flood, B.E.; Jones, D.S.; Bailey, J.V. *Sedimenticola thiotaurini* sp. nov., a sulfur-oxidizing bacterium isolated from salt marsh sediments, and emended descriptions of the genus *Sedimenticola* and *Sedimenticola selenatireducens*. *Int. J. Syst. Evol. Microbiol.* **2015**, *65*, 2522–2530. [CrossRef]

107. Cytryn, E.; van Rijn, J.; Schramm, A.; Gieseke, A.; de Beer, D.; Minz, D. Identification of bacteria potentially responsible for oxic and anoxic sulfide oxidation in biofilters of a recirculating mariculture system. *Appl. Environ. Microbiol.* **2005**, *71*, 6134–6141. [CrossRef]

108. Bengtsson, J.; Eriksson, K.M.; Hartmann, M.; Wang, Z.; Shenoy, B.D.; Grelet, G.A.; Abarenkov, K.; Petri, A.; Rosenblad, M.A.; Nilsson, R.H. Metaxa: A software tool for automated detection and discrimination among ribosomal small subunit (12S/16S/18S) sequences of archaea, bacteria, eukaryotes, mitochondria, and chloroplasts in metagenomes and environmental sequencing datasets. *Antonie Leeuwenhoek* **2011**, *100*, 471. [CrossRef]

109. Valle, A.; Fernández, M.; Ramírez, M.; Rovira, R.; Gabriel, D.; Cantero, D. A comparative study of eubacterial communities by PCR-DGGE fingerprints in anoxic and aerobic biotrickling filters used for biogas desulfurization. *Bioprocess Biosyst. Eng.* **2018**, *41*, 1165–1175. [CrossRef]

110. Sorokin, D.Y.; Tourova, T.P.; Bezsoudnova, E.Y.; Pol, A.; Muyzer, G. Denitrification in a binary culture and thiocyanate metabolism in *Thiohalophilus thiocyanoxidans* gen. nov. sp. nov.—A moderately halophilic chemolithoautotrophic sulfur-oxidizing *Gammaproteobacterium* from hypersaline lakes. *Arch. Microbiol.* **2007**, *187*, 441–450. [CrossRef]

111. Brito, J.; Valle, A.; Almenglo, F.; Ramírez, M.; Cantero, D. Progressive change from nitrate to nitrite as the electron acceptor for the oxidation of H_2S under feedback control in an anoxic biotrickling filter. *Biochem. Eng. J.* **2018**, *139*, 154–161. [CrossRef]

MDPI

St. Alban-Anlage 66

4052 Basel

Switzerland

Tel. +41 61 683 77 34

Fax +41 61 302 89 18

www.mdpi.com

ChemEngineering Editorial Office

E-mail: chemengineering@mdpi.com

www.mdpi.com/journal/chemengineering